D1352700

DIY robotics and sensors

on the commodore computer

practical projects for control application

john billingsley

First published 1984 by:
Sunshine Books (an imprint of Scot Press Ltd.)
12–13 Little Newport Street,
London WC2R 3LD

Copyright © John Billingsley, 1984

All rights reserved. No part of this publication may be reproduced, stored in a retrieval system, or transmitted in any form or by any means, electronic, mechanical, photocopying, recording and/or otherwise, without the prior written permission of the Publishers.

British Library Cataloguing in Publication Data
Billingsley, John
   DIY robotics and sensors on the Commodore computer.
   1. Robotics   2. Commodore computers
   I. Title
   629.8'92   TJ211

ISBN 0 946408 30 0

Cover design by Graphic Design Ltd.
Illustration by Stuart Hughes.
Typeset and printed in England by Commercial Colour Press, London E7.

CONTENTS

Contents in detail

CHAPTER 8
Interfacing a Robot
Building the robot language, teach mode, robot anatomy, six-axis control circuit.

CHAPTER 9
Analogue Output and Position Servos
Feedback and stability, circuits, analogue output from CB2 (PET only), interrupts, radio-control servos.

CHAPTER 10
Simple Robot Vision
Adding a vision system to a robot, two vision strategies, raster scan to screen, edge-following strategy.

CHAPTER 11
Whatever Next?
Robot evolution, intelligence, robot ping-pong contest.

CHAPTER 1
Getting Started

As the proud possessor of a Commodore microcomputer, you have more computing power at your disposal than served the whole of Cambridge University in 1953. You have no doubt played endless games, written programs of your own and explored many of the mysteries of the machine itself. Now what?

Until now, your machine has been dependent on keyed inputs for its performance — numbers on which to perform calculations or keystrokes to control games manoeuvres. Why not now let it get its own data? Why not add a muscle or two in the form of motors and relays, so that it can respond to the outside world? A new world of possibilities opens up, starting with turtles and robots, ending only at the bounds of imagination.

The simplest sensor channel uses one of the analogue-to-digital converter ports of the 64. Even on the PET, however, it is possible to measure an input voltage with no more than a resistor, a capacitor and the use of one bit of the user port — plus some cunning machine-code. Switches are simple to sense, and solid-state relays are little problem. Only when you want a variable output voltage do you have to consider adding more than a resistor or two — and even here it is possible to 'cheat' and obtain 256 output levels with two more resistors and another capacitor.

A lot can be achieved without a very deep technical knowledge of computer architecture, although I hope that you will learn about the fundamentals as you work through the book. As any new term is introduced it is put in quotes, and an attempt made to explain it. Now and again, quotes may be used for a 'buzz-word', for which an explanation is not really necessary.

The first few chapters may seem very simple and obvious. Often, however, it is the most trivial point, such as 'which way up does the connector go', which will catch you in the back of the neck. The robot chapter may seem rather specific to one brand of robot: in fact, it addresses the general problem of commanding a multitude of channels through a limited user port.

As each chapter was written, the designs and programs were tried and tested, and for this I am very grateful for the collaboration of my son, Richard, and of Timothy Dadd. Anything which has reached print should have worked at least once! Good luck.

Basic equipment

The components required for the construction of interfaces and systems are described in each chapter. In addition you will need a small soldering iron and some multicore solder. Reliable soldering is an art which can take years to perfect. One essential is to 'tin' both wires to be joined by melting fresh solder against each one separately. Never carry a blob of molten solder on the iron — in just a few seconds it can form a crust which will disguise a 'dry joint' and cause hours of troubleshooting.

You will also need a test meter. Buy a simple moving-coil multimeter, not a digital meter. You will be more interested in the rough value of a signal, whether it is ground or logic high, rather than its value to three places of decimals. Moreover, it is more convincing to see a needle move, requiring tens of microamps, than to see digits flicker which can be influenced by a small charge of static. The purpose of the meter is to eliminate uncertainty when something unexpected is happening, and it is little help if you must first wonder if the meter is showing a true story.

Although an attempt has been made to detail the components down to the last piece of wire, some extra wire will certainly come in useful — a metre or two of every colour you can find. Single strand 0.6 mm equipment wire will probably be easiest to use.

Construction methods

You will note on flicking through the pages that there is not a printed circuit in sight. When you have developed and proved a device, be it turtle or simple smoothing circuit, you may wish to construct a permanent version of the circuitry, laid out with great neatness. However this book is not concerned with knitting patterns. Instead it tries to establish the principles of interfacing gadgetry to a computer with a minimum of fuss. A three-dimensional rat's nest of components hanging on a connector strip may look untidy, but if carefully soldered the chances of error can be slight. Start with a lash-up which works, and only then transfer it to a more elegant form. Then when you find that the circuit no longer works, you will look for dry joints, hair-line cracks in the printed circuit tracks, or whiskers of copper bridging tracks which you thought you had cut.

Power supply unit

Some of the later designs call for a power supply capable of driving small motors. It is inadvisable to draw more than 100 milliamps from the computer, and so you will have to use a separate supply.

The components consist of a transformer with two 6V outputs, a diode bridge and two reservoir capacitors. This will give outputs of +7V and −7V, which can alternatively be used as a single 14V supply. You will also

need mains cable, at least one and preferably three fuses, three screw termi-
nals or three ways of connector strip, and a suitable box to mount it all in.

A 50 VA transformer will give plenty of margin against overload; the RS
207–245 should cost not much more than £5.00. A suitable 4 amp diode
bridge is the RS 262–113 costing a pound or so, whilst 2,200 microfarad
16V RS 103–373 capacitors are about fifty pence each. For the applications here, there is no need to smooth the supplies, but a single-package
stabilizer can easily be added later. The mains supply should be fused at 0.5
amp, and there should preferably be 3 amp fuses in the outputs.

What about the Vic 20?

The book is aimed principally at the Commodore 64 and the more recent
PET series, and the programs are carefully detailed to take care of the
differences between these machines. Many of the programs will run
equally well on the VIC 20, and also on the ancient 2001 PET. To list every
version of the programs in full would make them tedious, but, by substituting numbers from the tables below, you should be able to adapt the programs without too much trouble.

Terms such as 'port address' and 'data direction register' are dealt with
in some detail in Chapter 4. For now, the important thing is to remember
that this information is here, so that you can turn back to it when you need
it.

User port address:

All PETs: 59471 = $E84F   CBM 64: 56577 = $DD01   VIC 20: 37136 = $9110

Data direction register:

All PETs: 59459 = $E843   CBM 64: 56579 = $DD03   VIC 20: 37138 = $9112

Pointers to 'start of BASIC' and 'start of variables':

PET 30xx,
40xx,80xx: s.o.BASIC: 40, 41 = $28, $29   s.o.vars: 42, 43 = $2A, $2B
PET 2001:   s.o.BASIC: 122,123 = $7A, $7B   s.o.vars: 124,125 = $7C, $7D
CBM 64:   s.o.BASIC: 43, 44 = $2B, $2C   s.o.vars: 45, 46 = $2D, $2E
VIC 20:   s.o.BASIC: 43, 44 = $2B, $2C   s.o.vars: 45, 46 = $2D, $2E

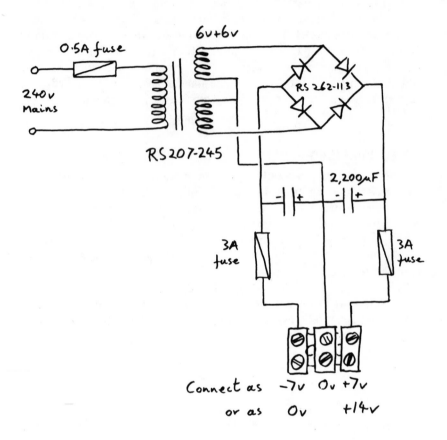

Figure 1.1   Simple power supply

BASIC program area starts at:

All PETs: 1025 = $0401   CBM 64: 2049 = $0801   VIC 20: 4097 = $1001

IRQ interrupt vector:

PET 30xx,40xx,80xx:	144,	145	= $ 90,	$ 91
PET 2001:	537,	538	= $219,	$21A
CBM 64:	788,	789	= $314,	$315
VIC 20:	788,	789	= $314,	$315

Program listings

The program listings given at the end of most of the chapters in this book draw together all the lines and routines listed in the course of each chapter. They are intended as a checklist, leaving out the REM statements.

CHAPTER 2
Signal Inputs

The computer thrives on a diet of numbers, stored in memory as binary digits or 'bits', and manipulated by the processor to form results which are also numbers. Within the computer electrical signals are either close to 5V, representing a logic 'one', or are close to ground and represent 'zero'. From combinations of such signals the numerical values are built up in the scale of two; one eight-bit byte can represent numbers from 0 to 255, two taken together can be interpreted as numbers from 0 to 65535, or alternatively from -32768 to $+32767$.

The outside world is not often like that. Keys on the keyboard do take binary values, either 'pressed' or 'not pressed', but other quantities such as positions of robots, speeds of motors or voltages on control knobs can vary continuously over their range. Somehow these 'analogue' values must be turned into numbers, so that the computer can digest them.

In this chapter, you will meet the interface already built into the Commodore 64 which allows you to connect up to four analogue signals. The signals are designed to take the form of variable resistances, but it is not too difficult to read voltages too. You can build a simple cable and connector strip, which will come in useful time and time again for trying out input schemes with a minimum of effort; and you can add a joystick which will enable you to drive the graphics program of the next chapter. You will also find out how to add a 'light-pen' to your computer, enabling you to select an item from the screen simply by pointing at it.

Connectors for the analogue ports (64 only)

The 64 is equipped with two connectors to be used with joysticks and games paddles. In addition to reading logic signals, each port has two analogue inputs to service an analogue joystick, and a simple POKE command will switch the circuitry from reading the pins of one connector to reading those of the other. One of the connectors also has a light-pen connection, which certainly merits later investigation.

A good confidence-builder is the construction and use of a home-grown joystick. Later, this can be regarded as a simple piece of test equipment to trouble-shoot analogue inputs in general. Even with a device as simple as this, containing only two potentiometers, the chances of non-working are legion. The important thing is to proceed step by step, eliminating uncertainty as you go.

First make up a connector to one of the games sockets, control port 1. You will need a 9-pin D-type male plug, such as Farnell 140−820 or RS 466−179, a 9-way (or more) length of 'chocolate block' connector strip and enough wire to join them — either 30 cm or so of ribbon cable or an equivalent assortment of coloured wires. In this way, the connections will be brought to a convenient connector beside the keyboard, so that you can trouble-shoot hardware and software together.

It is worth 'unscrambling' the pins, so that the signals on the connector strip are in a sensible order. The wiring order then becomes:

Connector strip:

Signal:	PotX	PotY	+ 5V	0V	Joy0	Joy1	Joy2	Joy3	But/LPen

Plug:

Pin No:	9	5	7	8	1	2	3	4	6

After wiring the connector strip, plug it in and check it. The secret of successful electronic development is not to trust *anything*. If you are convinced that there is a signal at one end of a wire, *check* that it really appears at the other — otherwise your faith in the wire and its soldering may cost you hours of searching for a bug.

First use a simple multimeter (6V DC range, − ve to the 0V line) to check out the + 5V and 0V connections. Also check Joy0, Joy1, Joy2 and Joy3, which will be between 4 and 5V. The use of a simple needle-and-scale meter does much more to inspire confidence than a flickering reading on a digital meter, especially when inspecting signals which are changing.

Now it is time to check the analogue input connections. Enter the following program into the micro:

```
10  PRINT PEEK(54297),PEEK(54298)
20  GOTO 10
```

and run it. The screen should fill with two columns of values of 255. Now connect a wire to + 5V via a 100 ohm resistor, and touch the end on to the analogue inputs in turn. You will see the appropriate column of the screen display a value close to zero as each channel is touched — your confidence is growing.

Finally check out the (switch-type) joystick input bits Joy0 to Joy3, and the 'Fire-button' bit But/LPen. Enter the program:

```
10  PRINT PEEK(56321) AND 31
20  GOTO 10
```

8

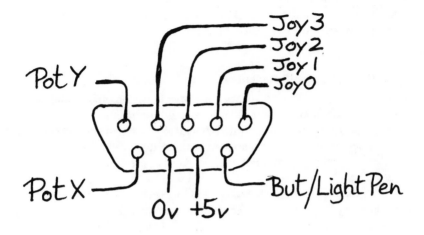

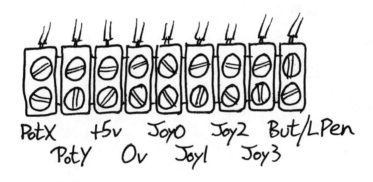

Figure 2.1   Analogue/games port connections

9

Remove the test wire from +5V and connect it instead to 0V (leaving out the resistor). While the program is running, touch the wire to each of the Joy pins and the But pin in turn. The value shown on the screen should change from its usual value of 31 to 30, 29, 27, 23 and 15 respectively. In other words, each bit will read '1' until it is connected to ground.

How does the analogue input work?

The principle of operation is very similar to the PET input which will be described in Chapter 5, but in the 64 the 'SID' sound chip does all the hard work. It has two input bits designed to read two analogue paddles, and the 64 has a separate switch to change over the connections from one pair of paddles to another. Each input channel works as follows:

The input bit is first connected to ground by the chip, discharging a capacitor. The capacitor is then allowed to charge up through the paddle's variable resistor, whilst the chip counts the time taken for the capacitor voltage to reach the trigger level. This count value is then transferred into another register inside the chip where it can be read with a simple PEEK. At switch-on, port 1 should be automatically selected. To make sure, you can

POKE 56322,PEEK(56322) OR 192

to set up the data direction register (see Chapter 4 for explanation). Then

POKE 56320,64

to select the paddle connections on port 1. If instead you want to select port 2, then you must

POKE 56320,128

— although this is not as straightforward as it seems. The chip tests the input several times per millisecond. Even so, if you change the switch within a program there should be a slight delay before the value is read. The values of the X and Y conversions are given by:

X = PEEK(54297)
Y = PEEK(54298)

The address which selects the switch position is also involved in reading the switch–joystick bits. Unfortunately it is also used as part of the keyboard reading system. This means that some fifty times per second an 'interrupt routine' will come blitzing through whatever else is happening, and will

10

leave the switch pointing to port 1 even if you really want port 2. For reading port 1 on its own there should be no problems in using BASIC, but, if you want to read both ports (ie all four channels), you must write a routine in machine code which inhibits interrupts for the vital moment.

You can now check out the port 2 connections by unplugging your test connector from one port and plugging it into the other. Now the Joy bits will appear at a different address, 56320. The test program is:

```
10 PRINT PEEK(56320) AND 31
20 GOTO 10
```

To test the analogue signals, use the following program. Lines 10 and 40 are simple two-instruction machine code programs. The instruction SEI is planted at 1024, which appears at the top left of the screen. This inhibits interrupts — which could be fatal if not corrected in line 40 by CLI. Both lines plant the code for a subroutine return at 1025, to bring control back to the BASIC program:

```
10 POKE 1024,7*16+8:POKE1025,6*16:SYS 1024:REM SEI RTS
20 POKE 56320,128
30 PRINT PEEK(54297),PEEK(54298)
40 POKE 1024,5*16+8:POKE1025,6*16:SYS 1024:REM CLI RTS
50 GOTO 10
```

Constructing a joystick

Now, how about the joystick. For this you will need two potentiometers of value 500 kohms. RS 161–830 would be suitable, but almost anything goes. Even the resistance value is not particularly critical; a value which is too low will restrict the range of resulting numbers, whilst if it is too high the useful region will be bunched up at one end of the rotation. Start simply. The potentiometer consists of a resistive track, connected to the outer solder tags. The middle solder tag is connected to the 'wiper', which slides along the track as the shaft is rotated and picks off a resistance value corresponding to its position. Connect one of the outer tags of one potentiometer to +5V. Connect the middle tag to PotX, then enter and run the first of the programs above. As you rotate the shaft to and fro, you will see the numbers in the first column vary from near 0 to 255.

They didn't? Then check the voltage on the potentiometer wiper using the multimeter. None there? Then take out the potentiometer and measure its resistance, and the resistance between the wiper and each end as the knob is rotated. Looks OK? Then put it back and try again; check the value of +5V. All looks OK, but still no changing numbers? Then check out PotX again as above; go to bed; try again in the morning.

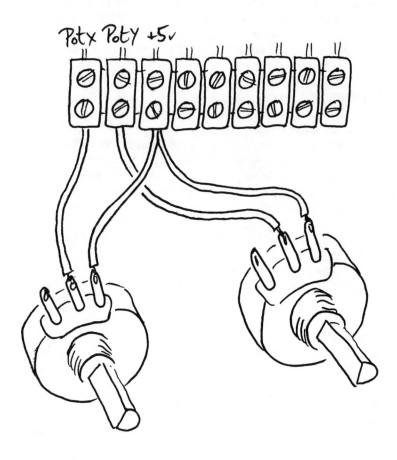

Figure 2.2   Connecting two potentiometers

Now connect one of the second potentiometer outer tags also to + 5V, and its wiper to PotY. When the test program is run, you should be able to control the numbers in both columns. Your electronic problems are at an end, and you are faced with the task of combining the mechanical movements into that of a joystick. One possibility is sketched in **Figure 2.3**.

If all else fails, a commercial joystick can be bought for under £10.00 — you might even prefer to buy an obsolete TV game and murder it to obtain a pair of joysticks.

So far, this chapter must have been most frustrating for owners of a PET. Even though the PET has no analogue input ports, it is still possible to attach a joystick — using a certain amount of skulduggery as described in Chapter 5. Before that can be done, it is necessary to understand the machinations of the user port, and the intricacies of the Versatile Interface Adaptor or VIA.

Adding a light-pen

Before we move on, remember that tempting light-pen connection which doubled as port 1's 'fire' button? What is a light-pen, and what can it do?

The television display is, of course, built up from a single spot which scans across the picture 15,000 times per second, working from top to bottom 50 times per second. If a pen containing a phototransistor is held against the screen, then as the spot passes beneath it, the phototransistor will give a pulse of output current. From the timing of this pulse, it is possible to work out the position on the screen which is being selected.

With the right interface, it should be possible to put a set of options on the screen — perhaps 'up', 'down', 'left', 'right' for robot commands — and to select an option merely by pointing to the word on the screen with a pen connected by a wire to the computer.

The 'official' light-pen interface costs £20.00 or so. Can you get away with anything simpler? A circuit consisting of one phototransistor OP500, three resistors and an NPN transistor 2N3705 is all you need — total cost under £1.00! If you turn up the TV brightness, you can make do with just the phototransistor and a single resistor.

The LPen connection leads to the VIC Video Interface Chip. This is an industrious device which turns a collection of numbers in memory into an appropriate picture on the screen. Suppose that the first line of text to appear on the screen is 'the quick brown fox'. The chip first looks at the memory location to be represented at the start of the line — here it holds the code for the letter 't'. From another part of memory it must now look up the shape of the letter 't', or especially it must look up the pattern for the top scan-line of the 't'. As this is being used to modulate the tube's electron beam, the chip picks up the code for 'h' and looks up the pattern for its top line, and so on to the end of the row of letters. When the scan line is

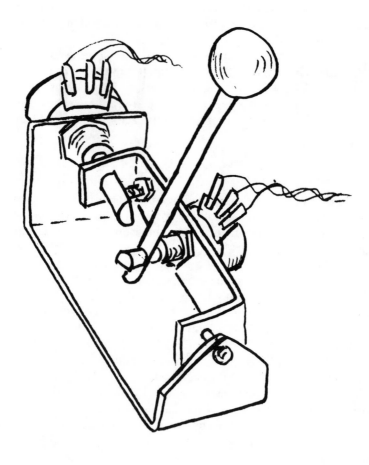

Figure 2.3 A simple joystick

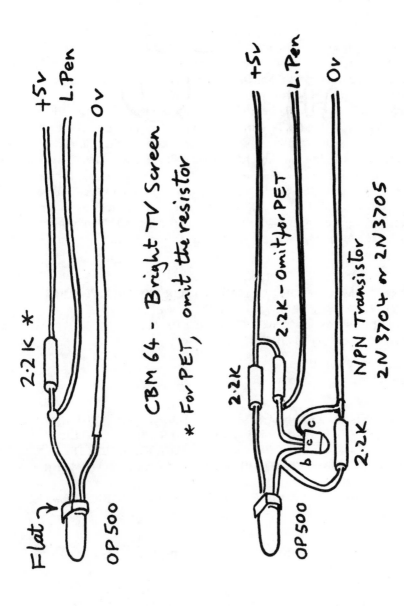

Figure 2.4   Light-pen for dim TV screen

complete, the chip looks at the 't' a second time, now looking up the pattern for its second line, and so on until eight scan lines have been output. The address of the letter being processed is held in a sixteen-bit register within the chip, and this is automatically incremented to follow the text down the page.

Up to this point, the description of the chip could fit the VIC chip or the CRT controller chip used in 8000 series and 'Fat 40' PETs. The CRT controller also has a light-pen connection, but its connection is brought out to pin 21 of connector J4 via a 7404 inverter with a 1 kohm pull-up. This connector is the memory-expansion forest of pins at the righthand side of the computer — you will have to be rather adventurous to use it. Whenever the LPen pin of the CRT controller changes from logic level 0 to 1, the address of the character being displayed at that instant is snapped by two 8-bit registers within the chip. From their value, it is possible to work out where the character appeared on the screen, and hence react appropriately. The chip is one of those awkward ones which try to conserve addressing space (see Chapter 4) by having to be prodded with the register number at one address, and parting with the answer at a second. Don't worry too much if the following program seems to be mumbo jumbo; it's short and it works:

```
10  REM SIMPLE LIGHT-PEN DEMO FOR 8000 OR FAT40 PET
20  REM FILL SCREEN WITH WHITE (GREEN)
30  FOR I = 32768 TO 34768: POKE I,160:NEXT
40  AO = 32768 : REM OLD HIT
```

Here comes the mumbo jumbo:

```
 50  CR = 14*16^3 + 8*16^2 + 8*16:REM ADDRESS OF CHIP IS $E880
100  POKE CR,17:A = PEEK(CR + 1): REM READ REGISTER 17
110  POKE CR,16:A = PEEK(CR + 1) + 256*A: REM REGISTER 16
120  B = A − 32768: REM B IS NOW CHARACTER NUMBER ON
     SCREEN
130  IF B< 0 OR B> 1999 THEN 100:REM MISSED SCREEN!(999 FOR
     FAT40)
```

Now let's use the result to move a screen blob:

```
140  POKE AO,160 :     REM RUB OUT OLD BLOB
150  POKE A,32 :       REM PLANT NEW BLACK BLOB
160  AO = A :          REM REMEMBER WHERE IT IS
170  GOTO 100:         REM AND DO IT AGAIN
```

16

Now let's get back to the 64 and its VIC chip. This chip is not content with handling colours, bit-mapped graphics and eight sprites, it also insists on giving the light-pen hit accurate to the nearest pair of pixels. There are now two registers at 53267 and 53268 which hold the X and Y values of the light-pen hit. The Y-value gives the exact scan-line number, whilst the X-value points to a pair of pixel positions. Since these are measured from the edge of the picture, behind the border, it is necessary to subtract 46 from each number to hit the display area. These values are strobed just once per frame scan, so you shouldn't move too fast.

The following program is just a rough-and-ready indication that the technique works. It doesn't take much imagination to extend it to drawing lines just two pixels thick, but for now let's be content with moving blobs.

```
10 REM SIMPLE LIGHT-PEN DEMO FOR COMMODORE 64
20 REM FILL SCREEN WITH WHITE
30 SC = 1024:CO = 55296 : REM SCREEN START, COLOUR
40 FOR I = 0 TO 999 : POKE I + SC,160
50 POKE I + CO,1:NEXT : REM COLOUR WHITE
60 AO = 0 : REM OLD HIT
70 PX = 53267:PY = 53268: REM ADDRESSES OF HIT REGISTERS
100 X = PEEK(PX) – 46:Y = PEEK(PY) – 46:REM READ REGISTERS
110 A = INT(X/4) + 40*INT(Y/8): REM WHICH CHARACTER?
120 IF A< 0 OR A> 999 THEN 100 : REM MISSED THE SCREEN!
```

Now let's plant a red blob:

```
140 POKE CO + AO,1 :   REM RUB OUT OLD BLOB — MAKE
                       WHITE
150 POKE CO + A,2 :    REM PLANT NEW RED BLOB
160 AO = A :           REM REMEMBER WHERE IT IS
170 GOTO 100:          REM AND DO IT AGAIN
```

This is a good simple program, however Tim and Richard felt that you would like something more entertaining. Below is their answer to the Stylophone. Doh, Re, Mi, etc., will appear on the screen, with a column of blobs to the left. Point the light-pen at a note, and that note will start to play; the column to the left gives blessed silence. Have pity on the neighbours.

Light-pen Demo

```
100 INPUT"SPEED (100)ISH";SP
110 GOTO10000
200 N=INT((PEEK(VY)-26)/16)
210 IF N<0 OR PEEK(VX)< 80 THEN N=0:GOTO200
220 POKEPH,PH(N):POKEPL,PL(N)
230 POKEW,0:POKEW,WW
```

```
240 FOR I=1 TO 2000 STEP SP:NEXT
250 GOTO 200
10000 RV$=CHR$(18):RO$=CHR$(146):REM REVERSE,OFF
10010 NN$="DO   RE   MIFA   SO   LA   TIDO"
10020 PRINTCHR$(147)CHR$(5);
10030 FOR I=1 TO 13: PRINT : PRINT"   "RV$"   "
10040 PRINT"   "RV$"   "RO$"                          ";
10050 IFI=2ORI=4ORI=7ORI=9ORI=11THENPRINT"        ";
10060 PRINTRV$"   "MID$(NN$,2*I-1,2);"        ";
10070 NEXT
10100 DIMPH(16),PL(16)
10110 P=2^(1/12):PO=4000
10120 FOR I=1 TO 13
10130 PH(I)=INT(PO/256):PL(I)=PO AND 255
10140 PO=PO*P:NEXT
10200 POKE 54296,15:POKE54277,7:W=54276:WW=17
10210 PH=54273:PL=54272
10220 VX=13*16^3+19:VY=VX+1
10230 POKE53281,0:GOTO200
```

CHAPTER 3
Graphic Design with a Joystick

I am married to a graphic designer. One day my wife, Ros, wanted to find out more about computer graphics, and to design some screen displays.

Graphic subroutines are all but non-existent on the Commodore, and, although the 64 has some powerful hardware graphics capabilities, the task of writing a full colour graphics package for it is not a trivial one. Unless you buy (or write) a machine-code package, the routines will be slow to execute. Much of the merit of the program developed here is in the clues it gives you for writing your own routines. All the same, I hope you will agree that it provides a useful set of functions, and is fun to drive. It makes a good excuse to use a joystick, and even includes a touch of digital filtering to remove the joystick's jitter. Alternatively, you can drive it from the light-pen developed in the previous chapter. Unfortunately graphics on the PET are only supported by an add-on board, and so this chapter is devoted to the 64.

Graphics and sprites

The VIC chip which controls the 64's display can operate in normal mode, displaying 1000 locations of memory, in the form of characters looked up in a table of shapes. Alternatively, it can display a 320 by 200 bit-image picture of 4000 memory locations. In both cases, a separate area of 1000 bytes of memory holds the foreground and background colour specification of each 'characters-worth' of display, organised as 40 cells across by 25 deep.

In addition, the chip will handle eight 'sprites'. These are bit-image shapes which can be superimposed on text or graphics, and can be moved bodily by changing only two locations each. Each sprite is defined by a sequence of 63 memory locations, giving a resolution of 24 bits across by 21 down — the sprites are linked to the shape definitions by pointers, and several or all of the sprites can be linked to the same shape. The following program is about as simple as you can get, and allows you to move a sprite around the screen by means of the joystick. It will establish a few principles, but needs a lot of embellishment to make it more interesting.

```
10  VV = 13*16^3:      REM VIC CHIP IS AT $D000
20  POKE VV+21,1:      REM TURN ON SPRITE 0 (BIT 0)
30  POKE 2040,13:      REM POINT SPRITE 0 TO
                         ADDRESS 64*13
40  POKE VV,160:       REM X OF SPRITE 0
50  POKE VV+1,100:     REM Y OF SPRITE 0
60  FOR I = 832 TO 832 + 62: POKE I,255:NEXT
70  REM FILL SHAPE WITH SOLID COLOUR
```

If you enter and run the program so far, you will see the sprite appear and fill up with colour. Now you will want to move it around the screen:

```
100  POKE VV,PEEK(54297):      REM JOYSTICK X
110  POKE VV+1,PEEK(54298): REM JOYSTICK Y
120  GOTO 100
```

And that's it! Other sprites are turned on by the remaining bits of $VV + 21$, and the shape pointer for sprite N is at $2040 + N$. The colour of sprite N is changed by poking $VV + 39 + N$ with a number from 0 to 15, whilst the X and Y values are at $VV + 2*N$ and $VV + 2*N + 1$ — with bit N of $VV + 16$ giving a boost of 256 to the X coordinate. Bits N of $VV + 23$ and $VV + 29$ will double the width and height of each corresponding sprite.

To design a sprite can be a tedious task. At the end of the chapter a 'sprite editor' is given which enables you to drive a pixel cursor around the sprite, setting and clearing bits at will.

Now let us attack the much more daunting task of building a graphics package to simplify the design of a complete screenful of coloured lines and shapes.

Program specification

The functions which the program provides are Point, Line and Area, whilst ink and paper colour can be set by pressing C. As the joystick is moved, a fleeting dot moves about the screen. Pressing P marks a fixed dot on to the screen, and also memorises the coordinates of the point.

If the joystick is moved and L is pressed, a line is drawn from the last recorded point. Another move and another L draws a second line from the end of the first, and so on. If the L key is held down, line segments will be drawn in swift succession, forming a smooth curve drawn by the joystick movement.

Areas are filled by shading them horizontally. A sets the boundary in the same way that L draws a line. If a boundary is already set at that vertical coordinate, then the line between the old boundary and the new is filled in with colour. If you hold down A and trace out a 'U' with the joystick, you will see the downstroke drawn as a curve on the screen. On the upstroke, the curve will 'fill up' like a wineglass. Whenever P is pressed, the old boundary will be forgotten.

To avoid accidentally erasing this work of art, the 'clear' command is an exclamation mark, requiring you to hold 'shift' at the same time.

Pressing C will return the screen to normal mode, and you will be asked for the new foreground and background colours. After you have entered them, the screen will change back to graphics mode, with your picture intact.

Setting up graphics mode

Specifying the program requirements is a good start, but how do we obtain high-resolution graphics mode on the 64? (I am afraid that PET owners are completely left out of this chapter.) We must send a few bytes of data to the VIC chip which controls the display, and we must also change a byte in CIA2. The next chapter attempts to explain some of the magic, but for now just take it on trust.

You may prefer to skip this paragraph the first time you read the book, and return to it later.

Bit image mode is set by poking register 17 of the VIC chip with a value which is ORed with 32. CIA2 at address $DD00 sets the 'bank address' for the VIC chip's display and colour memory, using the least significant bits. These are inverted, so that if their value is 3, the bank starts at $0000, 2 gives $4000, 1 gives $8000 and 0 gives $C000. Register 24 of the VIC chip defines the rest of the address with value 16* (colour offset) + (screen offset). These offsets are multiplied by 1024 to give the actual addition to the address. Thus, by selecting bank 1, and by poking VIC register 24 with value 16*7 + 8, we define the screen map to run from $6000 to $7FFF and the colour map to run from $5C00 to $5FFF. (Of course this restricts the RAM available for programs. The machine would alternatively work well using the 'RAM underneath the ROM', with screen at $E000 and colour at $C000: select bank 3, and POKE register 24 to value 8. All is well as you write to the display, but, as soon as you want to PEEK it, you will see the ROM instead. It's an easy problem to bypass in machine code, but cannot be done in BASIC — hence the use of bank 1.)

Now read on. Start building the program in testable modules. Enter the following program and run it. The 'housekeeping' is parked out of the way at 10000 onwards, whilst subroutines at 9000 and 9100 respectively set and reset graphics mode. If all is well, you will be asked for 'colour, pattern'. Try 33, 15 for a start. The screen will flip into high-resolution mode, and the colours will rapidly be set to black-and-white. The program will then nibble a pattern of stripes, taking quite a few seconds to cover the entire screen, finally flipping back to normal mode. If you break the program, typing GOSUB 9100 might bring rescue — otherwise try RESTORE.

```
    5 POKE 56,64:CLR: REM PROTECT GRAPHIC MEMORY
      AREA
   10 GOTO 10000

10000 REM
10020 SC = 1*16384 + 8*1024: REM SCREEN STARTS BANK 1,
      SECTOR 8
10025 CC = 1*16384 + 7*1024: REM COLOUR MAP = BANK 1,
      SECTOR 7
10030 VV = 13*4096:CI = VV + 13*256: REM VIDEO CHIP = $D000,
      CIA2 = $DD00
10040 GF = VV + 17:CM = PEEK(GF):GM = CM OR 32
10045 REM GRAPHICS FLAG, CHAR. MODE, GRAPHICS MODE.
10200 GOTO 100

  100 INPUT"COLOUR, PATTERN";CL,CH
  110 GOSUB 9000: REM SET GRAPHICS MODE
  120 FOR I = CC TO CC + 1000:POKE I,CL:NEXT:REM SET
      COLOUR
  130 FOR I = SC TO SC + 8000:POKE I,CH:NEXT:REM PATTERN
  140 GOSUB 9100: REM RESET TO CHARACTER MODE
  190 STOP

 9000 POKE CI,PEEK(CI)AND(255 – 1):REM SET BANK 1
 9010 POKE VV + 24,8 + 16*7:REM SCREEN SECTOR = 8, COLOUR
      = 7
 9020 POKE GF,GM:REM GRAPHICS MODE
 9030 RETURN

 9100 POKE CI,PEEK(CI)OR 3:REM BANK 0
 9110 POKE VV + 24,20:REM NORMAL SCREEN
 9120 POKE GF,CM:REM CHARACTER MODE
 9130 RETURN
```

As you will have guessed, the strange numbering will fit in with later additions.

Setting a point

The arrangement of the screen display memory is not altogether straight-forward, and it requires careful planning to translate an X,Y coordinate into the byte address and pattern to output. The first eight bytes form the rows of the first character position on the screen — they are displayed one below the other. The next byte is back at the top of the screen in the second character position, and so on. At the end of the top text line are bytes 39*8 to 39*8 + 7, followed by byte 40*8 at the top left of the second character line.

First let us calculate the character position on the screen. For each increase of 8 in Y (measured downwards) we will move down one row of 40 characters, ie we start with LL*INT(Y/8), where LineLength has been set to 40. Next, for each increase of 8 in X, we will advance one character, adding INT(X/8). The corresponding byte of the colour map may need to be poked with the colour, and checking the character position against zero and the maximum value of 999, tells us if the point is off the screen. Let us put the 'set a point' routine at 8500:

8500 CS = LL*INT(Y/8) + INT(X/8):   REM   CHARACTER POSITION
8510 IF CS< 0 OR CS> MX THEN RETURN: REM   MISSED   THE SCREEN

Now the byte address is found by adding 8*CS plus (Y AND 7) to the screen start address SC to give:

8520 SS = SC + 8*CS + (Y AND 7)

Now we must poke the byte at SS with a value which is ORed with the bit to be planted. This bit is in turn given by bit (X AND 7), which can be calculated as $2^{(7 - (X \text{ AND } 7))}$. To save calculation time, the values are held in an array BP(7), to give:

8530 POKE SS,PEEK(SS) OR BP(X AND 7)

Check to see if the character being written to is a new one, and, if so, set the colour:

8540 IF CS< > OS THEN POKE CC + CS,CL:OS = CS

then

8550 RETURN

We still have some housekeeping to attend to:

10000 Y = 0:X = 0:CS = 0:SS = 0:DIM BP(7)
10010 LL = 40:MX = 999

and

10050 FOR I = 0 TO 7:BP(I) = 2^(7 − I):NEXT

You can try out the routine up to this point by altering line 140 to:

140 FOR A = 0 TO 6.3 STEP .01

with

150 X = INT(100 + 90*COS(A)):Y = INT(100 + 90*SIN(A))
160 GOSUB 8500:NEXT
170 GOSUB 9100 :REM SET NORMAL SCREEN
100 CL = 33:CH = 0

and after the screen is cleared, a circle should appear.

Plotting straight lines

An important part of any graphics package is the ability to draw an oblique straight line from XH,YH to XT,YT (Here to There). We will come across similar routines later, when we program robots to move obliquely. Let us put this routine at 8000.

```
8000  DX = XT − XH:DY = YT − YH: REM  CALCULATE  SIZE  OF
                                MOVE
8010  IF ABS(DY)> ABS(DX) THEN 8100:   REM Y MOVE IS THE
                                       GREATER
8020  IF ABS(DX)< 1 THEN RETURN:    REM  NO  LINE,  GO
                                    HOME
8030  Y = YH:RA = DY/ABS(DX):       REM SLOPE OF LINE
8040  FOR X = XH TO XT STEP SGN(DX): REM FROM  HERE  TO
                                     THERE
8050  GOSUB 8500:Y = Y + RA:NEXT
8060  XH = XT:YH = YT:RETURN:        REM  HERE  IS  NOW
                                     THERE

8100  IF ABS(DY)< 1 THEN RETURN:     REM NO LINE
8110  X = XH:RA = DX/ABS(DY)
8120  FOR Y = YH TO YT STEP SIGN (DY)
8130  GOSUB 8500:X = X + RA:NEXT
8140  XH = XT:YH = YT:RETURN
```

To try out this part, change lines 140 to 160 to:

```
140  XH = 190:YH = 100:FOR A = 0 TO 50 STEP 2.2
150  XT = INT(100 + 90*COS(A)):YT = INT(100 + 90*SIN(A))
160  GOSUB 8000:NEXT
```

The shape filler

We need to be able to fill a line with colour. Let us put a subroutine at 9500, called after setting variables X and Y. If the shape memory SH(Y) is set, then the line will be filled from SH(Y) to X. If not, indicated by SH(Y) = − 1, then SH(Y) is set equal to X and the program returns.

Extra housekeeping is needed in the form of:

```
10060  M8 = 255:XM = 319:YM = 199:SH = 0: REM MAX X,Y SHAPE
         FLAG
10070  JX = 54297:JY = 54298:REM ANALOGUE ADDRESSES (FOR
         LATER)
10080  DIM SH(200),FL(7),FR(7):REM SHAPE, ROWS OF BITS
10090  FOR I = 0 TO 7:FL(I) = 2*BP(I) − 1:FR(I) = BP(I) − 1:NEXT
```

and the routine becomes:

```
8600  IF Y< 0 OR Y> YM OR X< 0 OR X> XM THEN RETURN
8610  X1 = X:X2 = SH(Y):SH(Y) = X:REM  SET  SHAPE  TO  NEW
        POINT
8620  IF X2< 0 THEN 8500:REM SHAPE WAS NOT SET, JUST DRAW
        EDGE
8630  IF X1> X2 THEN X1 = X2:X2 = X:REM SWAP
8640  CS = LL*INT (Y/8) + INT(X1/8):POKE CC + CS,CL:REM
        COLOUR
8650  SS = SC + 8*CS + (Y AND 7)
8660  I = (INT(X2/8) − INT(X1/8))*8:REM I = 0 SAME BYTE, 8 IF
        NEIGHBRS
8670  IF I = 0 THEN POKE SS,PEEK(SS)OR(FL(X1AND7) − FR(X2
        AND7)):RETURN
8680  POKE SS,PEEK(SS)OR FL(X1 AND 7)):REM LH END OF LINE
8690  POKE  SS + I,PEEK(SS + I)OR  (M8 − FR(X2  AND  7)):REM  RH
        END
8700  POKE CC + CS + I/8,CL
8710  IF I = 8 THEN RETURN:REM NO MIDDLE TO FILL
8720  X1 = SS + 8:X2 = SS + I − 8:CS = CC + CS + 1:FOR SS = X1 TO X2
        STEP 8
8730  POKE SS,M8:POKE CS,CL:CS = CS + 1
8740  NEXT:RETURN
```

We are also going to need a routine to reset the shape array to -1, and to reset a flag:

```
9600 IF SH> 0 THEN FOR I = 0 TO YM:SH(I) = - 1:NEXT
9610 SH = 0:RETURN
```

Now for the routine which calls the line filler. This looks similar to the line-drawing routine, but operates just once for each value of Y:

```
8200 DX = XT - XH:X = XH
8210 IF  ABS(YT - YH)< 1  THEN  XH = XT:X = XT:Y = YH:GOSUB
     8600: RETURN
8220 R = DX/ABS(YT - YH)
8230 FOR Y = YH TO YT STEP SGN(YT - YH)
8240 GOSUB 8600:X = X + R:NEXT
8250 XH = XT:YH = YT:RETURN
```

Change 160 to

```
160  GOSUB 8200:NEXT
```

to try out the new section.

The joystick routine

So far, we have developed routines which can be called from within a program to produce graphics on the screen. How do we link them to movements of the joystick? We need a joystick routine which will read the two analogue values, and then translate them into movements on the screen in the range 0 to 320 for X and 0 to 200 for Y. Since the analogue value is an integer in the range 0 to 255, we have the choice of losing resolution in the X direction, or losing the righthand fifth of the screen — the latter seems the better bet. There must then be a dot shown in the selected position, which must be removed when the joystick is moved to a new place. The dot will show up best if it twinkles. This means that we must first restore the byte of screen memory at the old joystick position, remembered in variable OB (old byte), and plant a new byte in the new position with the appropriate bit reversed in value. Then we return to go round again.

```
5000 POKE SJ,OB: REM PUT BACK OLD BYTE
5010 XT = PEEK(JX):REM READ ANALOGUES
5020 YT = PEEK(JY)
5030 CS = LL*INT(YT/8) + INT(XT/8)
```

```
5040  SJ = SC + 8*CS + (YT  AND  7):REM  ADDRESS  OF  SCREEN
      BYTE
5050  OB = PEEK(SJ)
5060  I = BP(XT AND 7):POKE SJ, (OB OR I) – (OB AND I):REM FLIP
      BIT
5070  RETURN
```

When you come to try the program, you will probably find that the selected spot is subject to a certain amount of jitter. It is easy to include digital filtering in the program to reduce this. Consider first the simple instruction:

$$X = PEEK(JX)$$

As soon as the analogue value changes, the value of X will change to match it. Now consider instead:

$$X = X + (PEEK(JX) - X)/2$$

If the PEEK value has been zero for a while, and suddenly changes to 100, then the next value of X will be $0 + (100 - 0)/2 = 50$. The following value will be $50 + (100 - 50)/2 = 75$, and so on. Each time through the program the difference between X and the PEEK value will halve, so that X will eventually catch up with PEEK, although the effect of sudden changes will be smoothed out. Using a number bigger than 2 in the program line will give more smoothing, but X will take longer to catch up with the PEEK. This smoothing system is called a 'low pass filter'. It gives the same effect that would be gained by putting a series resistor and shunt capacitor into the analogue circuit. If the program line is made:

$$X = X + (PEEK(JX) - X)/F$$

the value of F can be chosen to give a variety of time-constants. The longer the time-constant, the less effect jitter noise will have, but the slower will be the response of the joystick.

Now, if filtering is desired, lines 5010 and 5020 can be replaced by

```
5010  XT = XT + (PEEK(JX) – XT)/F
5020  YT = YT + (PEEK(JY) – YT)/F
```

The rest of the program

Now let us deal with the remainder of the housekeeping. We must first clear the screen and initialise the values of OB and SJ. Life will be easier if we move the clear-screen routine to 9700 from its 'try-out' position at 110:

```
9700  FOR I = CC TO CC + 999:POKEI,CL:NEXT:REM CLEAR
      COLOUR
9710  FOR I = SC TO SC + 7999:POKEI,0:NEXT:REM CLEAR
      DISPLAY
9720  FOR I = SC TO SC + 3:POKEI,M8:NEXT
9730  RETURN:REM SET TOP LEFT CHAR TO SHOW SET
      COLOUR
```

We can now add:

```
10100  CL = 22:GOSUB 9000:GOSUB 9700:REM WHITE ON BLUE
10110  F = 4:GOSUB 5010:REM JOYSTICK SET-UP
10120  GOTO 100
```

Now we are ready for the main loop of the program.
    First read the joystick, showing its position as a dot:

```
100  GOSUB 5000
```

Next test for a key-press. If none, loop via the joystick test. The key-press test is a little unusual, because we would like the user to be able to hold down a key to repeat an operation. For this type of command, automatic repeat can be a menace, and it is better to look at the key-code echo at location 197; this will have value 64 if no key is pressed. If a key is held down, then GET B$ will give the string value of the key the first time and " " subsequently. Thus the following code should do the trick:

```
110  GET B$
120  IF B$< > "" OR PEEK(197) = 64 THEN A$ = B$
```

If no key is pressed, go round the loop again:

```
130  IF A$ = "" THEN 100
```

Is the key a 'P'? If so, set the point, clear the shape (if set) and loop:

```
140  IF A$< > "P" THEN 170
150  OB = OB OR BP(XT AND 7):GOSUB 9600
160  XH = XT:YH = YT:A$ = "":GOTO 100
```

Is the command 'L'? If so, draw the line and loop:

```
170  IF A$ = "L" THEN GOSUB 8000:GOTO 100
```

If the command is 'Area' then call the area routine, if it is '!' then clear the screen:

180 IF A$ = "A" THEN SH = 1:GOSUB 8200:GOTO100
190 IF A$ = "!" THEN GOSUB 9700:GOTO 100

Otherwise the command is 'C' — or it is not recognised. In either case, revert to normal screen:

200 GOSUB 9100: IF A$< > "C" THEN 240
210 PRINT "FOREGROUND COLOUR (0 – 15), BACKGROUND":
     INPUT I,CL
220 GOSUB 9000:CL = (CL + 16*I)AND M8
230 POKE CC,CL:GOTO100:REM SHOW AT TOP LEFT

If not recognised, print a message and wait for another key:

240 PRINT CHR$(147)"P – POINT   L – LINE"
250 PRINT "A – AREA   C – COLOUR"
260 PRINT "! – CLEAR SCREEN"
270 GET A$:IF A$ = "" THEN 270
280 GOSUB 9000:A$ = "":GOTO100

Now you can let your artistic talents run wild. You will need a very steady hand to drive the joystick when holding down a key for continuous writing. The results can be most impressive — especially if you are married to a graphic designer.

Graphics by Joystick

```
5 POKE 56,64:CLR:REM PROTECT GRAPHIC MEMORY AREA
10 GOTO 10000
100 GOSUB 5000
110 GET B$
120 IF B$<>"" OR PEEK(KB)=KO THEN A$=B$
130 IF A$="" THEN 100
140 IF A$<>"P"THEN 170
150 OB=OB OR BP(XT AND 7): GOSUB 9600
160 XH=XT:YH=YT:A$="":GOTO 100
170 IF A$="L" THEN GOSUB 8000: GOTO 100
180 IF A$="A" THEN GOSUB 8200: GOTO 100
190 IF A$="!" THEN GOSUB 9700: GOTO 100
200 GOSUB 9100: IF A$<>"C" THEN 240
210 PRINT "FOREGROUND COLOUR (0-15), BACKGROUND"
```

```
215 INPUT I,CL
220 GOSUB 9000: CL=(CL+16*I)AND M8
230 POKE CC,CL: GOTO 100:    REM SHOW AT TOP LEFT
240 PRINT CHR$(147)"P-POINT      L-LINE"
250 PRINT "A-AREA       C-COLOUR"
260 PRINT "!-CLEAR SCREEN"
270 GETA$: IFA$="" THEN 270
280 GOSUB 9000:A$="":GOTO 100
5000 POKE SJ,OB:              REM PUT BACK OLD BYTE
5010 XT=XT+(PEEK(JX)-XT)/F
5020 YT=YT+(PEEK(JY)-YT)/F
5030 CS=LL*INT(YT/8)+INT(XT/8)
5040 SJ=SC+8*CS+(YT AND 7)
5050 OB=PEEK(SJ)
5060 I=BP(XT AND 7):POKE SJ,(OB OR I)-(OB AND I)
5070 RETURN
8000 DX=XT-XH: DY=YT-YH
8010 IF ABS(DY) > ABS(DX) THEN 8100
8020 IF ABS(DX)<1 THEN RETURN
8030 Y=YH: RA=DY/ABS(DX)
8040 FOR X=XH TO XT STEP SGN(DX)
8050 GOSUB 8500: Y=Y+RA: NEXT
8060 XH=XT: YH=YT: RETURN
8100 IF ABS(DY)<1 THEN RETURN
8110 X=XH: RA=DX/ABS(DY)
8120 FOR Y=YH TO YT STEP SGN(DY)
8130 GOSUB 8500: X=X+RA: NEXT
8140 XH=XT: YH=YT: RETURN
8200 DX=XT-XH:X=XH:XH=XT
8210 IF YT=YH THEN X=XT:Y=YH: GOSUB 8600: RETURN
8220 R=DX/ABS(YT-YH)
8230 FOR Y=YH TO YT STEP SGN(YT-YH)
8240 GOSUB 8600: X=X+R: NEXT
8250 XH=XT: YH=YT: RETURN
8500 CS=LL*INT(Y/8) + INT(X/8)
8510 IF CS<0 OR CS>MX THEN RETURN
8520 SS=SC+8*CS+(Y AND 7)
8530 POKE SS,PEEK(SS) OR BP(X AND 7)
8540 IF CS<>OS THEN POKECC+CS,CL: OS=CS
8550 RETURN
8600 IF Y<0 OR Y>YM OR X<0 OR X>XM THEN RETURN
8610 X1=X: X2=SH(Y): SH(Y)=X
8620 IF X2<0 THEN 8500
8630 IF X1>X2 THEN X1=X2:X2=X
8640 CS=LL*INT(Y/8)+INT(X1/8):POKE CC+CS,CL
8650 SS=SC+8*CS+(Y AND 7)
8660 I=(INT(X2/8)-INT(X1/8))*8:IF I=0 THEN 8750
8680 POKE SS,PEEK(SS)ORFL(X1 AND 7)
8690 POKE SS+I,PEEK(SS+I)OR(M8-FR(X2 AND 7))
```

```
8700 POKE CC+CS+I/8,CL
8710 IF I=8 THEN RETURN
8720 X1=SS+8:X2=SS+I-8:CS=CC+CS+1
8725 FORSS=X1 TO X2 STEP 8
8730 POKE SS,M8: POKE CS,CL:CS=CC+1
8740 NEXT: RETURN
8750 POKE SS,PEEK(SS)OR(FL(X1 AND7)-FR(X2 AND7))
8760 RETURN
9000 POKE CI,PEEK(CI) AND (255-1):REM SET BANK 1
9010 POKE VV+24,8+16*7
9020 POKE GF,GM:              REM GRAPHICS MODE
9030 RETURN
9100 POKE CI,PEEK(CI) OR 3:          REM SET BANK 0
9110 POKE VV+24,20
9120 POKE GF,CM:              REM CHARACTER MODE
9130 RETURN
9600 IF SH>0 THEN FOR I=0 TO YM: SH(I)=-1: NEXT
9610 SH=0: RETURN
9700 FOR I=CC TO CC+999: POKE I,CL: NEXT
9710 FOR I=SC TO SC+7999:POKE I,0: NEXT
9720 FOR I=SC TO SC+3: POKE I,M8: NEXT
9730 RETURN: REM TOP LEFT CHARACTER SHOWS COLOUR
10000 Y=0: X=0: CS=0: SS=0: DIM BP(7)
10010 LL=40: MX=999:KB=197:KO=64
10020 SC=1*16384+8*1024:REM SCREEN=BANK1 SECTOR8
10025 CC=1*16384+7*1024:REM COLOUR=BANK1 SECTOR7
10030 VV=13*4096:CI=VV+13*256:REM VIDEO CHIP,CIA
10040 GF=VV+17: CM=PEEK(GF): GM=CM OR 32
10045 REM GRAPHICS FLAG,CHAR. MODE,GRAPHICS MODE
10050 FOR I=0 TO 7: BP(I)=2^(7-I): NEXT
10060 M8=255:XM=319:YM=199:SH=0:      REM MAX X,Y
10070 JX=54297:JY=54298:     REM ANALOGUE ADDRESS
10080 DIM SH(200),FL(7),FR(7)
10090 FOR I=0 TO   7
10095 FL(I)=2*BP(I)-1: FR(I)=BP(I)-1:NEXT
10100 CL=22:GOSUB 9000:GOSUB 9700:REM WHITE/BLUE
10110 F=8: GOSUB 5010
10120 SH=1:GOSUB9600
10200 GOTO 100
```

Sprite Editor

```
10 GOTO2000:      REM SPRITE EDITOR STARTS AT 2000
200 X=160:Y=100:DX=0:DY=0:X2=0:Y2=0:REM FUN DEMO
210 A=20:B=300:AY=40:BY=230
220 M=255:V4=V+4:V5=V+5:VH=V+16
```

31

```
300  X2=.1*(RND(1)-.5):Y2=.4*(RND(1)-.3)
310  X1=X1+X2:Y1=Y1+Y2
320  IF X<A THEN X1=ABS(X1)
325  IF X>B THEN X1=-ABS(X1)
330  IF Y<AY THEN Y1=ABS(Y1)
335  IF Y>BY THEN Y1=-ABS(Y1)
340  X=X+X1:Y=Y+Y1
350  POKE V5,Y:POKE V4,X AND M:POKE VH,-4*(X>M)
360  GOTO 300
2000 V=53248:POKEV+21,4:POKE2042,13
2005 SB=832:POKEV+4,100:POKEV+5,100
2010 POKEV+23,4:POKEV+29,4:PRINT CHR$(147);
2020 E=0:B=7
2030 U$=CHR$(145):D$=CHR$(17):   REM CURSOR CHARS
2040 L$=CHR$(157):R$=CHR$(29)
2050 PRINT"SPACE TO CLEAR, . TO SET"
2060 PRINT"CURSOR TO MOVE, X TO END"
2100 P=PEEK(SB+E):C=2^B:Q=(P OR C)-C
2200 POKE SB+E,Q:FORI=1TO10:NEXT:POKE SB+E,Q ORC
2210 GET A$:IFA$=""THEN:FORI=1TO10:NEXT:GOTO2200
2220 POKE SB+E,P
2230 IFA$=" "THEN P=Q:A$=R$:POKE SB+E,P
2240 IFA$="."THEN P=Q OR C:A$=R$:POKE SB+E,P
2250 IFA$=R$ THEN B=B-1:IFB>=0 THEN 2100
2260 IFA$=R$ THEN B=7:E=E+1:IFE>62 THEN E=0
2270 IFA$=L$ THEN B=B+1:IF B<8 THEN 2100
2280 IFA$=L$ THEN B=0:E=E-1:IF E<0 THEN E=63
2290 IFA$=D$ THEN E=E+3:IF E>62 THEN E=E-63
2300 IFA$=U$ THEN E=E-3:IF E<0 THEN E=E+63
2310 IFA$="X"THEN PRINT CHR$(147):GOTO 200
2400 GOTO 2100
```

CHAPTER 4
Logic In and Out

In Chapter 2, some mention was made of the internal goings-on of the computer and the difference between analogue and digital signals. Even when the external signals are digital, bringing them to the computer's attention is not altogether an easy matter.

Digital interfaces

The spine of the microcomputer is made up of two 'buses', bunches of signals which link most of the components together. The simplest of these is the data bus. When the processor (another name for the microprocessor chip) wants to store a data byte in memory, it switches the corresponding logic voltages on to the eight-bit data bus, issues a command, and the appropriate memory location remembers the data. When the processor wants to retrieve a byte, whether of data or the next instruction in its program, it sends out its command; the memory looks up the data and applies a corresponding set of logic voltages to the data bus for the processor to read. You will see that the data bus is extremely busy, and an attempt to apply external signals to it could well send the processor diving in a spin.

To order the memory about, the processor must be able to specify an address. Now we come across the second bus, the sixteen-bit address bus. 65,536 different addresses can be specified as the bus stands, but a little cheating can extend it to address any size of memory you can afford. Another important line allows the processor to tell the memory whether it wants to read or write data.

What has all this to do with interfacing? Clearly, something has to be placed between any logic input line and the data bus, so that the data is only allowed on to the bus for the brief instant when the processor wants to read it. This is the role of the interface chip. It means, for instance, that eight pins of the chip can be connected to a plug on the machine for the convenience of the user. Slip a socket on to this connector, and you can attach extra keys, sensor contacts, or any other sort of logic signal, and, with a program command or two, read them into the computer. This then is the user port.

Making connections to the user port

Before getting to grips with the user port, it is a good idea to bring it within reach. This entails making a connecting cable, similar to the one described in Chapter 2, to bring the signals to a connector strip beside the keyboard. The connector in this case is a 12-way double-sided edge connector providing 24 contacts with .156 inch spacing. Suitable codes are Cinch 251.12.90.160, Amp 530657−3 and Teka TP3−121−E04. The connections of the logic data lines are the same for both the 64 and the PET, but the + 5V line to be found on the 64's connector is missing from the PET. PET users must use a second connector attached to one of the cassette ports (against the advice in the PET manuals — but OK up to 50 mA).

The connector pins are (looking in towards the computer):

1	2	3	4	5	6	7	8	9	10	11	12
A	B	C	D	E	F	H	J	K	L	M	N

Strip:	1	2	3	4	5	6	7	8	9	10	11	12
CBM 64:	2	B	C	D	E	F	H	J	K	L	M	N
Signal:	+5V	Flg	PB0	PB1	PB2	PB3	PB4	PB5	PB6	PB7	PA2	Gnd
Pet:	*	B	C	D	E	F	H	J	K	L	M	N
Signal:	+5V	CA1	PA0	PA1	PA2	PA3	PA4	PA5	PA6	PA7	CB2	Gnd

* (taken from PET cassette port pin B or 2)

After plugging the edge connector into the computer, check out the signals using the multimeter. Connect the negative test lead to position 12. Check for + 5V on position 1 — if it is not there, perhaps the socket is the wrong way round. Signals PB0 to PB7 (or PA0 to 7) — from now on let us call them P0 to P7 — can be checked by a mystic means which will become clear later in this chapter. Enter the program:

```
10  PO = 56577 : REM : *** CBM 64
or
10  PO = 59471 : REM : *** PET
20  PRINT 255 − PEEK(PO)
30  GOTO 20
```

Run the program, and connect a wire from 0V in turn to P0, P1 up to P7. The values 1, 2, 4, 8, 16, 32, 64 and 128 should appear on the screen respectively.

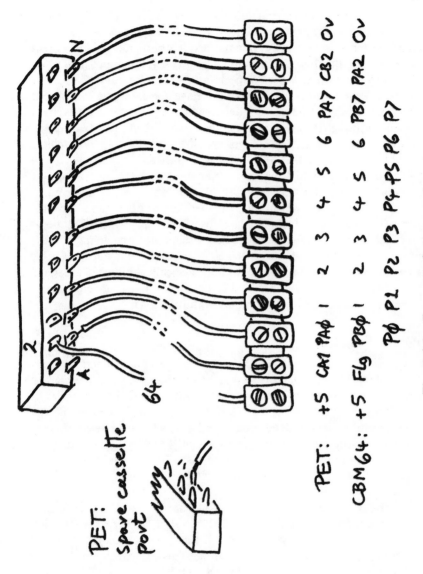

Figure 4.1   Connector for user port

How the interface works

Now at last we are in a position to look at the operation of the user port. The signals which concern us at first are P0 to P7, which are taken from the B-port of a Complex Interface Adaptor in the case of a 64, and from the A-port of a Versatile Interface Adaptor in the case of the PET. The PET's VIA is so versatile that it can baffle in an instant. It is a single chip, a 6522, memory-mapped to appear at address $E840 (the $ denotes hexadecimal). If this makes sense to you, skip the next few paragraphs. Not content with being versatile, the 6526 CIA chip is complex. It is mapped at $DD00.

Early computer systems used a special bus system for controlling input /output, and many microcomputers still have special input and output instructions. It was soon realised, however, that inputs and outputs could be treated as though they were memory locations. When a number is saved in memory, say in location $1234, voltages are altered within the circuitry of one of the memory chips. If these were amplified and connected to the outside world, they could drive eight output lines, so that storing a value of zero would set all eight lines low, whilst 255 would set them all high. When the contents of a memory location are loaded, on the other hand, the logic values of voltages stored in one of the chips will be copied into the accumulator of the microcomputer. Suppose that instead of stored voltages, these signals came from wires connected to eight voltages in the outside world, then we would have eight inputs. An interface chip can thus be designed which will have a lot in common with a memory chip — and some manufacturers produce chips which combine both functions.

Address decoding

The sixteen address lines of the 6502 microcomputer can directly address 65536 separate bytes of memory — many more than will fit on the average memory chip. The top few address bits are thus decoded to give lines to address individual memory chips, whilst the remaining bits are connected to all chips in parallel to determine the address within the selected chip. (This is not absolutely true for some sorts of RAM, but ignore that for now.) Not all the memory chips need to be present to make the system work, so there may be blank spaces within the memory map of the machine.

If the top four lines are decoded, there will be sixteen 'chip-select' lines, the first responding to addresses from $0000 to $0FFF, the next from $1000 to $1FFF and so on. One of these, the $D line for instance, could enable another decoder to decode the next four lines, giving sixteen more signals which would respond to addresses $D0XX, $D1XX ..... $DFXX, where XX can be any two hex digits. One of these lines, say the one which responds to $DDXX, can enable yet another decoder, giving a further sixteen lines which respond to $DD0X, $DD1X, etc. Finally, one of these lines, say, the $DD0X line, could enable a chip with just 16 memory addresses, $DD00 to $DD0F. Now suppose that, instead of being a genuine memory chip, this chip can be connected to the outside world. Then, if we save the value 7 (binary 00000111) in address $DD01, we can control eight output lines to make three pins go high and another five pins go low. That, in a nutshell (coconut?), is the principle of memory-mapped input/output. The trouble is that we now have to crack open the nut — and the 6522 and 6526 are hard nuts to crack!

Ports and data direction registers

Sixteen bytes have 128 bits. If we had separate lines for inputs and outputs that would leave us with an awful lot of pins on the chip. After providing the signals required to connect the chip into the micro system (18 lines) plus two lines for power supply, a 40-pin pack does not have much to spare. The 20 remaining pins are arranged as two ports, each with eight data lines and two control or 'handshake' lines. It is one of these ports which should by now be connected to the connector strip beside your keyboard. It may look like a connector to you, but the computer is convinced that it is the memory byte at the address $DD01 (64) or $E84F (PET). Each port is bi-directional, that is each individual bit can be an input or an output. The direction of each bit is held in a register within the chip called (wait for it) the Data Direction Register. For the user port, the 64 sees DDR-B at $DD03 whilst the PET addresses DDR-A at $E843. Each bit which is made a 1 at this address will be selected as an output bit, whilst the 0s will select inputs.

Now, how do we go about looking at a specific address location? The expression PEEK(..) is used to denote 'contents of address'. Thus the command PRINT 5 will print the value 5, whilst PRINT PEEK(5) will print the contents of memory location 5. Similarly the command POKE 5,6 will save the value 6 in memory 5 — and may crash the system at the same time!

Now get ready with your multimeter again. Set all the user port bits to outputs by typing:

PO = 56577:DD = 56579 for the CBM 64
or
PO = 59471:DD = 59459 for the PET

then POKE DD,255
    Also type:

POKE PO,255

to set all the bits high. Now check with your meter, and see that P0, P1, etc., are all at about +5V. Now type

POKE P0,0

and see that P0, P1, etc., have all dropped to 0V. Now, using the values 1, 2, 4, 8, 16, 32, 64 and 128, ensure that you can set the lines high one at a time. Now try various combinations — see why it's easier to use hexadecimal?

Now try configuring the port as an input. Just type:

POKE DD,0

and every line will be an input. Enter the program:

```
10  PRINT PEEK(56577) : REM 64
```
or
```
10  PRINT PEEK(59471) : REM PET

20  GOTO 10
```

and run it. As you touch a 0V wire on to each pin P0, P1, etc., you will see the number change from 255. After a short battle of mental arithmetic, you may prefer to change the program to:

```
10  PRINT 255 – PEEK(56577) : REM OR 59471
20  GOTO 10
```

Sinks and sources

You have, of course, noticed that you have had to prod the input with 0V to change it, not 5V. Each pin has an internal pull-up resistor which, if left alone, will hold the input at 5V, and the computer will read it as a '1'. To pull the input down to a '0', the input signal must be able to 'sink' about 1mA of current. This current corresponds to one 'TTL load', and sets a limit on the number of logic inputs which a TTL logic gate output can drive. To have a good 'fan-out', a TTL gate must be able to sink several milliamps, but does not really need to 'source' any current at all to work. The pulling-down capability of the logic outputs is therefore much better than the pulling-up power. (Although the 6522 and 6526 are MOS devices, they are designed to be TTL compatible.)

More versatile yet!

Of course that is not the whole story about the 6522. Four of its sixteen addresses are taken up with data and data direction — that leaves twelve more on which to build its reputation for versatility. Two pairs of addresses concern two sixteen-bit counters, which can be used for a variety of timing functions. Another controls a shift register, used to transfer data in and out of CB2. A further register, the flag register, indicates the conditions which could have caused an interrupt, such as timer-expired, data-handshake, etc. Finally two more registers, the Peripheral Control Register and the Auxiliary Control Register orchestrate the whole variety show. Even the simple-looking port B has some surprises up its sleeve, for PB7 can produce a pulse of variable width or a pulse train under the control of timer 1, whilst PB6 can be used as an input for pulses to be counted by timer 2.

The 6526 is a close relative of the 6522, but tries to go one better. It has a built-in time-of-day clock, which counts cycles of mains at 50 or 60 cycles. Otherwise it is much the same, apart from a refinement to its serial output which completely messes up the pseudo-analogue output scheme described in Chapter 9 for the PET with its 6522. With the addition of two resistors and a capacitor, the PET can give a signal which is ideal for commanding an analogue servo-motor. The 64 has to find alternative methods.

Of greater immediate interest to PET owners, is how to input an analogue signal from a joystick, and this warrants a short chapter of its own.

Switching mains voltages

Having come to grips with the port, now is the time to put it to some use. A particularly useful device when power-switching is required is the 'solid-state-relay'. This is in fact an opto-isolated triac, which will switch an AC mains load of several amps on or off at will. Provided all necessary precautions are taken to avoid stray conductors (or, especially, fingers) bridging between the signal end and the 'hot' end of the device, the opto-isolation makes it a safe device for connecting to the user port. Connection could hardly be simpler; the + ve pin is connected to + 5V, whilst the − ve pin is attached to the P bit, which has been chosen to operate the unit. Whenever this bit is configured as an output, and when the corresponding output data bit is zero, the switch will be on.

The RS number of the 2.5 ampere device is RS 348−431. At a price pushing £10.00, you may not want to add too many channels. The simple program given below could be made to switch a reading lamp on and off at random times. With simple modifications, the times could be made less random to give the impression of somebody working late, then going to bed. With the addition of a photocell and a PEEK or two, the system can respond to fading daylight. Beware, however, that the passing burglar does not deduce that the lamp is flashing the presence of a computer in the house!

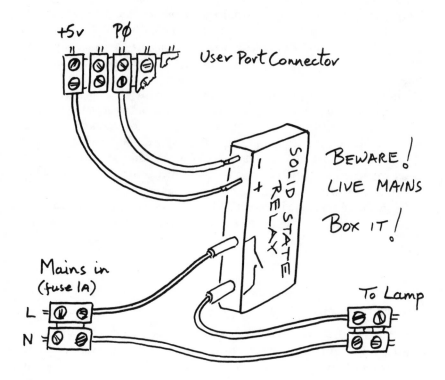

Figure 4.2   Solid state relay for light control

```
 10  PRINT CHR$(147)"TIMES IN MINUTES:"
 20  INPUT "MAX ON-TIME ";OM
 30  INPUT "MAX OFF-TIME ";FM
 40  INPUT "MIN ON-TIME ";OL
 50  INPUT "MIN OFF-TIME ";FL
 60  IF (OL-OM> 0) OR (FL-FM> 0) THEN GOTO 10

100  PO = 56577: DD = 56579 : REM *** CBM 64
```
or
```
100  PO = 59471: DD = 59459 : REM *** PET

110  POKE DD,1: REM MAKE BIT 0 AN OUTPUT

200  POKE PO,0: REM TURN ON LIGHT
210  T = OL + (OM-OL)*RND(1): REM BETWEEN OL,OM
220  GOSUB 1000

300  POKE PO,1: REM TURN OFF LIGHT
310  T = FL + (FM-FL)*RND(1)
320  GOSUB 1000
330  GOTO 200

1000  REM WAIT T MINUTES
1010  T = T*60 : REM MAKE SECONDS
1020  FOR I = 1 TO T
1030  FOR J = 1 TO 1000: NEXT REM ONE SECOND DELAY
1040  NEXT
1050  RETURN
```

It's a bit primitive as it stands, but I am sure that you will be able to add any extra features you need.

CHAPTER 5
Analogue Input for the PET and 64

PET owners will by now be impatient to know how to input a joystick paddle signal without needing to add an analogue-to-digital converter chip. The method described here uses no more than a single bit of the user port. The roots of the technique lie in the depths of antiquity — they are at least five years old. Although the same principle lies behind the converter built into the 64, it can be extended to give high-quality conversion suitable for instrumentation.

How it works

The old single-chip TV tennis games needed to encode the joystick signals with a minimum of resources. There was no room for an analogue-to-digital conversion — the chip would be hard pressed to process the digital value at display speeds anyway. Instead the joystick variable resistance was connected to a capacitor, giving a variable time-constant. Let us consider horizontal bat movement. At the start of each TV line the capacitor was discharged. As the line was then scanned, the capacitor charged up via the joystick resistance. As the capacitor passed the threshold voltage of the input connection, the screen dot brightened to write the image of the bat. In other words, the joystick resistance was converted into a delay, which could be read by a single input bit. Can't we play the same trick with one bit of the user port?

For demonstration purposes try a large value of capacitor first, so that the timing can be done with a loop of BASIC program. Afterwards reduce the capacitor so that machine code will give a swift answer. Start with 1000 microfarads (a 6V electrolytic is actually quite small). Connect this between 0V (– ve end) and P0. Now enter and run the following program:

```
10  PO = 59471: DD = 59459 :REM *** PET
```
or
```
10  PO = 57577: DD = 56579 :REM *** CBM 64

20  POKE DD,1:   REM configure PB0 as an output
30  POKE PO,0:   REM zero output to discharge capacitor
40  GOSUB 1000: REM brief delay
```

```
100  C = 0:          REM set count to zero
110  POKE DD,0:      REM PB0 becomes an input, capacitor released
120  IF (PEEK(PO) AND 1)> 0 THEN 200 :REM got to threshold ?
130  C = C + 1:      REM keep counting
140  GOTO 120:       REM round again
200  POKE DD,1:      REM discharge the capacitor again
210  PRINT C
220  GOSUB 1000:     REM brief delay
230  GOTO 100:       REM do it all again
1000 FOR I = 1 TO 200: NEXT: RETURN
```

The numbers which appear on the screen will depend on the exact value of the capacitor. Disconnect the capacitor and the numbers should fall to zero. Now connect a 2 kohm potentiometer, as a variable resistance in series with the capacitor, to P0 — ie capacitor + ve end to potentiometer wiper, one end of potentiometer to P0, — ve end of capacitor to 0V. While the program runs, the number printed should change as the potentiometer shaft is turned.

A high-speed machine code version

Some readers may be familiar with assembly language and machine code, while to others the subject might be a complete mystery. If, after reading the next few paragraphs, the subject is an even greater mystery, do not despair. The software is given in the form of BASIC data statements which you can enter in the usual way, and from then on you can rely on blind faith to make the program run. Your faith in your own typing ability should not be quite so blind! If you type the letter O instead of the number zero, the program will almost certainly crash and is likely to be totally lost. Therefore before trying to run it, save it.

To run at an acceptable speed, the routine must be rewritten in machine code (ie assembly language) and should ideally take the form of a USR function, so that you could include the line:

X = USR(CH)

within a BASIC program, to return the value of channel CH. This entails linking the machine code to the USR jump address, and also calling the floating-point-to-integer conversion routine (and vice versa). With a variety of PETs to deal with, not to mention the 64, all with routines in different places, I prefer to tackle the problem in a different way.

Variables in the Commodore machines all take the same form. In particular, the fourth byte (ie byte 3 — numbered from 0) of an integer variable will contain the least-significant byte of the value. Now, if that variable is

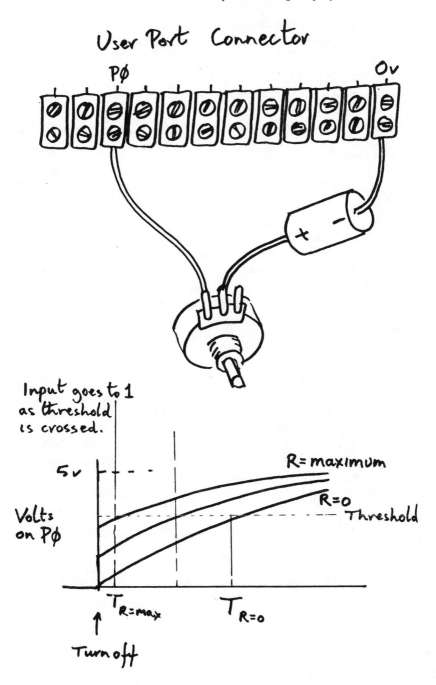

Figure 5.1   Simple analogue input

the very first one to be declared, its location will be pointed to by bytes 42, 43 (PET 30xx, 40xx, 80xx), or bytes 45, 46 (CBM 64) or bytes 124, 125 if you collect antique 2001s. Now two assembler instructions:

```
LDY    #$03
LDA    (VARS),Y
```

will load the accumulator with its value. Suppose the second variable to be declared is the integer V%, then:

```
LDY    #$0A
STA    (VARS),Y
```

can be used to store the result in V%.

The next problem is finding a safe place to tuck the machine code. An old favourite is the second cassette buffer, starting at location $033A = 826. This used to be safe in the old days, but, in 40xx PETs, it is now getting decidedly crowded with disk signals. Others prefer the top of memory, pulling the ceiling down to prevent the code being trampled by strings. My own preference is to make a sandwich in the BASIC, the bottom slice being at the normal BASIC start address, containing a line such as (for 40xx PETs):

```
10 POKE 41,12: RUN
```

Then follows a generous filling of machine code, topped off at $0C01 by the slice of BASIC which drives the machine code and takes the hard work out of displaying or filing the results. For demonstration purposes, however, where the program has to be loaded from data statements, we might as well stick with the cassette buffer, using a patch at $0390 which is relatively traffic-free.

First the machine code must be loaded. Let us put the loader up out of the way at 10000.

```
   10  CH% = 0:V% = 0:GOTO 10000

10000  MC = 3*16^3 + 9*16 : I = MC : REM $0390
10010  READ A$: IF LEN(A$)< > 2 THEN 100 :REM DONE IF XXXX
10020  GOSUB 10100
10030  POKE I,A: PRINT I,A$,A
10040  I = I + 1: GOTO 10010

10100  A = ASC(A$) – 48 + 7*(A$> ":"): REM CONVERT FROM HEX
10110  B$ = MID$(A$,2)
10120  A = 16*A + ASC(B$) – 48 + 7*(B$> ":")
10130  RETURN
```

Now we can write the assembler in a recognisable form consisting of data statements (of course you need not type in the REM parts). Do watch out for capital Os which should be zeros!

```
10200 DATA A0, 03     :REM           LDY #3          PET
10210 DATA B1, 2A     :REM           LDA (VARS),Y    ****
10220 DATA A8         :REM           TAY
10230 DATA 4D, 43, E8 :REM           EOR DDR         *
10240 DATA 8D, 43, E8 :REM           STA DDR         *
10250 DATA 98         :REM           TYA
10260 DATA A2, 00     :REM           LDX #0
10270 DATA 78         :REM           SEI
10280 DATA 2C, 4F, E8 :REM LOOP      BIT PORT        *
10290 DATA D0, 05     :REM           BNE DONE
10300 DATA E8         :REM           INX
10310 DATA D0, F8     :REM           BNE LOOP
10320 DATA A2, FF     :REM           LDX #$FF
10330 DATA 0D, 43, E8 :REM DONE      ORA DDR         *
10340 DATA 8D, 43, E8 :REM           STA DDR         *
10350 DATA 58         :REM           CLI
10360 DATA 8A         :REM           TXA
10370 DATA A0, 0A     :REM           LDY #10
10380 DATA 91, 2A     :REM           STA (VARS),Y    ****
10390 DATA A9, 00     :REM           LDA #0
10400 DATA 8D, 4F, E8 :REM           STA PORT        *
10410 DATA 60         :REM           RTS
10420 DATA XXXXX      :REM           END
```

This version will work with 30xx, 40xx and 80xx series PETs. The old 2001 PET has its pointers to the start of variables at addresses 124 and 125, and so the code contained in lines 10210 and 10380 will need changing to:

```
10210  DATA B1, FC
10380  DATA 91, FC
```

For the 64, the port and data direction addresses also need changing. There are enough changes to warrant listing this part of the program again:

```
10200 DATA A0, 03     :REM           LDY #3          CBM  64
10210 DATA B1, 2D     :REM           LDA (VARS),Y
10220 DATA A8         :REM           TAY
10230 DATA 4D, 03, DD :REM           EOR DDR
10240 DATA 8D, 03, DD :REM           STA DDR
10250 DATA 98         :REM           TYA
10260 DATA A2, 00     :REM           LDX #0
10270 DATA 78         :REM           SEI
10280 DATA 2C, 01, DD :REM LOOP      BIT PORT
```

```
10290 DATA  D0, 05       :REM              BNE DONE
10300 DATA  E8           :REM              INX
10310 DATA  D0, F8       :REM              BNE LOOP
10320 DATA  A2, FF       :REM              LDX #$FF
10330 DATA  0D, 03, DD:REM DONE            ORA DDR
10340 DATA  8D, 03, DD:REM                 STA DDR
10350 DATA  58           :REM              CLI
10360 DATA  8A           :REM              TXA
10370 DATA  A0, 0A       :REM              LDY #10
10380 DATA  91, 2D       :REM              STA (VARS),Y
10390 DATA  A9, 00       :REM              LDA #0
10400 DATA  8D, 01, DD:REM                 STA PORT
10410 DATA  60           :REM              RTS
10420 DATA  XXXXX    :REM                  END
```

Every so often the processor is interrupted, to go off and deal with some housekeeping such as reading the keyboard. In a BASIC program this is not noticeable, but, if it happens in the middle of the timing loop above, it will mess up the result. The command SEI blocks the interrupt — but always be sure to re-enable it with CLI afterwards, or take the consequences!

Now you can add a routine to test out the conversion. Remember that the channel number is set in CH%, with a result returned in V%. The actual conversion is carried out by a call SYS MC, where MC contains the address of the machine code.

```
100  CH% = 1:SYS MC:V1 = V%:  REM READ VALUE FROM P0
110  CH% = 2:SYS MC:V2 = V%:  REM READ VALUE FROM P1
120  PRINT V1,V2
130  GOTO 100 : REM MEASURE THEM AGAIN
```

Enter the program and set it running. With nothing attached to the user port, two columns of zeros should appear on the screen. Now connect a 2 microfarad capacitor in series with a 2 kohm potentiometer between P0 and 0V. You should find that, as you turn the potentiometer, you can obtain numbers in column 1 which vary from 0 to 255. If the maximum number is less than 255, use a correspondingly larger capacitor. If the smallest number is not 0, use a potentiometer of a higher value of resistance.

Now give P1 the same treatment, and you have the makings of a joystick. Of course, you can easily increase the number of input channels up to eight, one for each bit of the user port. There is not even any need to modify the program — it will cope with 8 channels as it stands, provided they are called as 1, 2, 4, 8, 16, etc.

The time for caution is when you have other inputs and outputs using the remaining bits of the user port. If you allow CH% to have the value 240 on calling SYS MC then the top four bits of the port will be left configured as outputs. The program as it stands will also output zeros to all output bits; as long as your own BASIC program ensures that the bits which are to be read as analogue inputs are set to zero, you can simply omit line 10400.

A more accurate converter

The analogue input provision on the 64, and the converter described so far in this chapter, are all very well for games joysticks, but they fall far short of being usable for serious applications where linearity might be required. They rely on the variation of a resistance for their input signal, and often a signal may instead be available in the form of an analogue voltage, say between 0 and 5 volts. This is the form which will be given as the output of an instrumentation amplifier, perhaps used for strain gauges or such.

The modification to obtain a linear voltage input is a relatively slight one, but still only gives an eight-bit result. For many applications this will be sufficient, but a further modification to the software would allow the precision to be as many bits as you like — although after twelve bits or so noise will tend to swamp any improvement. The penalty for increasing the resolution is an increase in conversion time. For eight bits the conversion takes about four milliseconds, and, at the very best, the conversion time will double for every extra bit which is added. A point is soon reached where it is preferable to perform the conversion in a special hardware chip — although now added bits can cost extra in money.

The hardware refinement to linearise the input involves adding an integrator and a comparator — two chips together costing under a pound. This will in fact give you four channels of analogue input, taking up five bits of the user port.

Now an extra bit is needed to control the integrator — we will use bit P4. When this is high, the output of the 747 operational amplifier (which forms the integrator) runs rapidly negative until caught at 0V by the diode. When P4 is pulled low, the integrator is allowed to run positive to reach +5V in a time 100K * .047 microfarads = $10^5 * .047 * 10^(-6) = 4.7$ milliseconds. The integrator ramp output is applied to the non-inverting input of each of the four comparators which make up an LM339. As the ramp passes each of the input signals, which are connected to the inverting inputs, the corresponding output bit will change from 0 to 1. This means that only a slight change to the software given above is needed.

First, the output of bit P4 must be pulled low at the start of conversion. This is achieved by adding a line to the software:

10235 DATA 09, 10 :REM ORA #$10

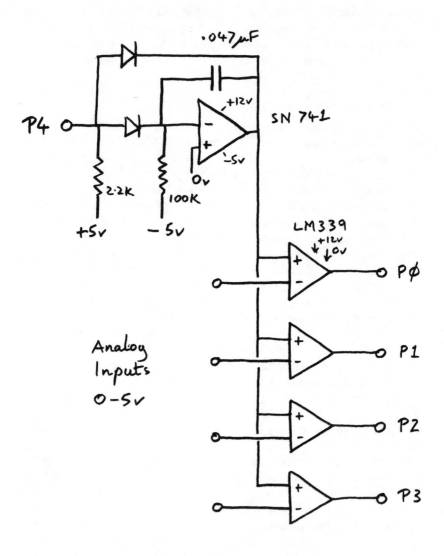

Figure 5.2   Linear analogue input

At the end of conversion the bit P4 must be allowed high again to reset the integrator to 0V. This is performed by a new line 10335:

10335 DATA 29, EF :REM AND #$EF

Now the software is compatible both with simple joysticks connected as before and with linear conversion of voltages in the range 0 to 5V using the interface of **Figure 5.2** — try it and see.

Laboratory instrumentation

Now that PET owners can read an analogue signal or a joystick, they will want to find an excuse to use it. The graphics system of Chapter 3 is out of their reach unless they have bought a high-resolution graphics system, but a much simpler use of graphics will enable a high-resolution chart to be plotted down the screen. This can enliven physics experiments by catching transients at up to fifty or so readings per second, or by delegating to the computer the task of sitting patiently taking readings every half hour.

To save the data points, it is necessary to open an array by substituting some new lines at 100 onwards:

```
100  POKE 59468,12:PRINT CHR$(128 + 14):REM SET GRAPHICS
     MODE
110  PRINT CHR$(147) :REM CLEAR SCREEN
120  INPUT "HOW MANY DATA POINTS";NP
130  INPUT "HOW MANY CHANNELS (1 TO 4)";NC:NC = NC - 1
140  DIM R(NP,NC),CC(NC)
150  FOR I = 0 TO NC:CC(I) = 2^I:NEXT:REM SET UP CHANNEL
     CODES
160  INPUT "DELAY BETWEEN SAMPLES (WHOLE SECONDS)"
     D
170  D$ = RIGHT$(STR$(1000000 + INT(D)),6):T0$ = "000000"
```

Now we can start to log data:

```
1000  FOR P = 1 TO NP:PRINT
1010  FOR C = 0 TO NC:CH% = CC(C) :SYS MC
1020  R(P,C) = V%: GOSUB 2000 :REM STORE DATA, PLOT IT
1030  NEXT C:GOSUB 1500:NEXT P:REM DELAY BETWEEN
      POINTS
1040  END
```

Subroutine 1500 must provide the timing function, waiting until the appropriate second has ticked:

```
1500  IF TI$< D$ THEN 1500
1510  TI$ = T0$:RETURN
```

We are now left with writing a display routine at 2000 onwards, capable of showing off the results to their best advantage. The very simplest method is to 'tab' across the screen with TAB(V%/7), and then to print a symbol corresponding to the channel number:

```
2000  PRINT CHR$(145) :REM CURSOR UP RETURN FOR LEFT
        MARGIN
2010  PRINT TAB(V%/7); CHR$(C + 49);
2020  RETURN
```

Try out the program as developed so far.

With a resolution of only forty points across the screen, this is not particularly inspiring. Let us get higher resolution by using the graphics characters, each representing a vertical line in one of eight positions — this will give 320 display values, and so none of the analogue resolution need be wasted. The graphics characters are defined by adding lines 180 onwards:

```
 180  DIM G$(7):FOR I = 0 TO 7
 190  READ C:G$(I) = CHR$(C):NEXT
10300  DATA 165,212,199,194,221,200,217,167
```

Now line 2010 can be replaced by

```
2010  PRINT TAB(V%/8);G$(V% AND 7)
```

Unfortunately, there is now no way of telling the channels apart, so, every 16 lines, the traces should be labelled. This can be achieved by putting back a variation on the old line 2010 as:

```
2005  IF (P AND 15) = 0 THEN PRINT TAB(V%/8);CHR$(C + 49)
        :RETURN
```

Having caught your data, you may want to display it again by typing a direct command GOTO 3000. The routine at 3000 to play it out will simply be:

```
3000  FOR P = 1 TO NP:PRINT:FOR C = 0 TO NC
3010  V% = R(P,C):GOSUB 2000:NEXT:NEXT
3020  END
```

Now try this final version. If you have lost track of its development, it is listed in full at the end of the chapter. Its structure is simple enough so that you should easily be able to modify it to analyse the data automatically, perhaps plotting the log or the square of one channel, and perhaps the ratio of some others. It is the starting point for completely automated experiments, where the computer will provide output signals to provoke the process under investigation, whilst observing and analysing the results.

Analogue Input for PET

```
10 CH%=0:V%=0:GOTO 10000: REM *** PET
100 CH%=1:SYSMC:V1=V%
110 CH%=2:SYSMC:V2=V%
120 PRINT V1,V2:GOTO 100
130 GOTO 120
10000 MC=3*16^2+9*16:I=MC
10010 READ A$:IF LEN(A$)<>2THEN100
10020 GOSUB 10100
10030 POKE I,A:PRINT I,A$,A
10040 I=I+1:GOTO 10010
10100 A=ASC(A$)-48+7*(A$>":")
10110 B$=MID$(A$,2)
10120 A=16*A+ASC(B$)-48+7*(B$>":")
10130 RETURN
10200 DATA A0,03, B1,2A, A8, 4D,43,E8
10210 DATA 8D,43,E8, 98, A2,00, 78
10220 DATA 2C,4F,E8, D0,05, E8, D0,F8
10230 DATA A2,FF, 0D,43,E8, 8D,43,E8
10240 DATA 58, 8A, A0,0A, 91,2A, A9,00
10250 DATA 8D,4F,E8, 60
10260 DATA XXXX
```

Analogue Input for CBM 64

```
10 CH%=0:V%=0:GOTO 10000: REM *** 64
100 CH%=1:SYSMC:V1=V%
110 CH%=2:SYSMC:V2=V%
120 PRINT V1,V2:GOTO 100
130 GOTO 120
10000 MC=3*16^2+9*16:I=MC
10010 READ A$:IF LEN(A$)<>2THEN100
10020 GOSUB 10100
10030 POKE I,A:PRINT I,A$,A
10040 I=I+1:GOTO 10010
10100 A=ASC(A$)-48+7*(A$>":")
10110 B$=MID$(A$,2)
10120 A=16*A+ASC(B$)-48+7*(B$>":")
10130 RETURN
10200 DATA A0,03, B1,2D, A8, 4D,03,DD
10210 DATA 8D,03,DD, 98, A2,00, 78
10220 DATA 2C,01,DD, D0,05, E8, D0,F8
10230 DATA A2,FF, 0D,03,DD, 8D,03,DD
10240 DATA 58, 8A, A0,0A, 91,2D, A9,00
10250 DATA 8D,01,DD, 60
10260 DATA XXXX
```

Data Acquisition and Plotting—PET

```
10 CH%=0:V%=0:GOTO 10000: REM *** PET
100 POKE 59468,12:PRINT CHR$(128+14)
110 PRINT CHR$(147)
120 INPUT"HOW MANY DATA POINTS";NP
130 INPUT"HOW MANY CHANNELS(1 TO 4)";NC:NC=NC-1
140 DIM R(NP,NC),CC(NC)
150 FOR I=0 TO NC:CC(I)=2^I:NEXT
160 INPUT"SAMPLING INTERVAL (WHOLE SECONDS)";D
170 D$=RIGHT$(STR$(1000000+INT(D)),6)
175 T0$="000000"
180 DIM G$(7):FOR I=0 TO 7
190 READ C:G$(I)=CHR$(C):NEXT
1000 FOR P=1 TO NP:PRINT
1010 FOR C=0 TO NC:CH%=CC(C):SYS MC
1020 R(P,C)=V%:GOSUB 2000
1030 NEXT C:GOSUB 1500:NEXT P
1040 END
1500 IF TI$<D$ THEN 1500
1510 TI$=T0$:RETURN
2000 PRINT CHR$(145);:IF (P AND 15)>0 THEN 2010
2005 PRINT TAB(V%/8);CHR$(C+49):RETURN
2010 PRINT TAB(V%/8);G$(V% AND 7):RETURN
3000 FOR P=1 TO NP:PRINT:FOR C=0 TO NC
3010 V%=R(P,C):GOSUB 2000:NEXT:NEXT
3020 END
10000 MC=3*16^2+9*16:I=MC
10010 READ A$:IF LEN(A$)<>2THEN100
10020 GOSUB 10100
10030 POKE I,A:PRINT I,A$,A
10040 I=I+1:GOTO 10010
10100 A=ASC(A$)-48+7*(A$>":")
10110 B$=MID$(A$,2)
10120 A=16*A+ASC(B$)-48+7*(B$>":")
10130 RETURN
10200 DATA A0,03, B1,2A, A8, 4D,43,E8
10210 DATA 8D,43,E8, 98, A2,00, 78
10220 DATA 2C,4F,E8, D0,05, E8, D0,F8
10230 DATA A2,FF, 0D,43,E8, 8D,43,E8
10240 DATA 58, 8A, A0,0A, 91,2A, A9,00
10250 DATA 8D,4F,E8, 60
10260 DATA XXXX
10300 DATA 165,212,199,194,221,200,217,167
```

CHAPTER 6
Stepper Motors and Their Use

Stepper motors are a favourite actuator for obtaining motor output. Their drives involve only logic signals, with no need for digital-to-analogue conversion. Until recently, only precision 'upper-class' motors were available at an outrageous price, but, with the microcomputer and a requirement for low-cost peripherals, there has come a demand for cheap stepper motors which the industry has been swift to fulfil. A suitable motor for turtles and micromice is the Philips ID35, distributed by Impex of Richmond at around £12.00.

Problems and principles

Despite their apparent advantages, stepper motors are not without their problems. They have a firm restriction on their top speed, and the useful torque falls off dramatically as this is approached. Sudden speed changes, even at relatively low speeds, can stall the motor. Unfortunately, unless special sensors are added, the computer is unaware that the motor has slipped 'out of cog'. All subsequent movements therefore take place with a position error, until a reset manoeuvre is made. Another drawback in a battery-driven system is power consumption; even when stationary, a stepper draws as much power as under full load.

Just how does a stepper motor work? The rotor is a permanent magnet, whilst the stator (the fixed case) has a number of electrical windings which when energised create a magnetic field. The field pulls the rotor into line, and, by changing the selection of energised windings in a suitable sequence, the rotor is pulled round step by step. When the stepping stops, the rotor is held in position by the magnetic field.

A simple stepper demonstrator

The movement of the permanent magnet rotor can be likened to the rotation of a magnetic compass — indeed you can use a compass in an experiment to demonstrate how a stepper motor operates. Obtain a cheap compass — the simple sort with a pointer rather than an ornate card will be best. Wind a coil of 50 turns of fine enamelled copper wire — 36 SWG or finer — across the compass. Obviously the wire must not obscure the view of the needle.

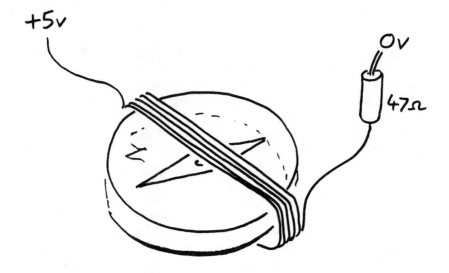

Figure 6.1   Compass and coil

Connect a 47 ohm resistor in series with the coil, and apply 5V across the ends. You can take 100 milliamps from the 5V pin on the user port connector — although this is the top limit. Of course, experimenting is made easy by using the chocolate block connector strip and cable described in Chapter 4.

When the voltage is applied, the needle should rotate and line up almost perpendicular to the coil, ie along the axis of the coil. Reverse the applied voltage, and the needle will reverse. Could the coil, and hence the needle, be driven directly from two bits of the user port? Unfortunately, the current available from PB0−7 is limited to about 3mA: unless you are prepared to wind coils of several hundred turns, this will not dominate the effect on the needle of the earth's magnetic field. The PET uses PA0−7, which have even less drive. We must therefore use some amplification — no bad thing in preparing to drive genuine stepper motors. The simplest amplifier consists of just one resistor and one transistor per bit of output — four of each per motor. (Later on, we can consider using a Darlington driver chip instead.) A good general purpose PNP transistor is a 2N 3703 (RS 294−334), costing well under £1.00 per pack of five. First connect just one transistor to your coil, driving it from PB0 via a 1 kohm resistor as in **Figure 6.2.**

Connect the circuit and switch on. Nothing should happen to the compass at first. Now tell the computer the addresses to use for the Data Direction register as variable DD and for the output POrt as variable PO by typing:

DD = 56579 :   PO = 56577 :   REM ***** CBM 64
or
DD = 59459 :   PO = 59471 :   REM ***** PET

Now you can set the output data register to all-bits-high by typing:

POKE PO,255

Next, configure bits 0−3 as outputs by typing:

POKE DD,15

Still nothing should happen, because the output of bit 0 is high, and does not yet sink any current via the transistor base. Now type:

POKE PO,255−1

This will take bit 0 to zero and current will flow into PB0 (or PA0 for PET) from +5V through the transistor base and R1. The transistor will be turned on, applying 5V from the transistor collector to the coil and

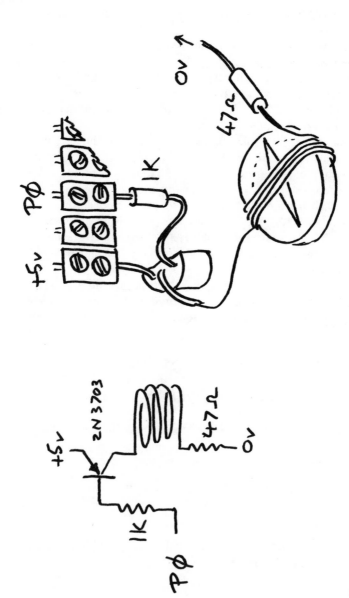

Figure 6.2   Transistor driver

resistor. The needle should leap into action. Turn the current off again
with:

POKE PO,255

before the resistor R2 starts cooking.

To reverse the needle, we must be able to pass current in the opposite
direction. With a circuit as simple as this one, we cannot reverse the current
in the wire, and so we need a second coil, wound directly over the top of the
first. Wind a further fifty turns of wire, connecting one end to the resistor,
and winding in a direction such that the two joined wires become the half-
way point of the coil which now has one hundred turns. Connect a twin
version of the transistor circuit, and drive it from user port bit 1.

Now the commands:

POKE PO,255−2

followed by:

POKE PO,255−1

should drive the compass needle first one way (North, say) and then the
other (South). Another command:

POKE PO,255

will switch off both arms of the coil, and the compass will be left to the
mercy of the earth's field.

So far, messing about with a compass does not seem to have much to do
with motors. But now the plot gets more exciting. Wind another twin coil,
also of 50 + 50 turns, over and perpendicular to the first coil, connecting
the mid-point (or 'centre-tap') of the new coil to the same 47 ohm resistor
and the centre-tap of the first coil. Now when the ends of the new coil are
connected to the collectors of two more transistors which are driven from
bit 2 and bit 3, the command:

POKE PO,255−4

will cause the needle to point in the new direction. If the first coil caused the
needle to point North or South, then the second coil causes the needle to
point East or West. By switching on one of the N−S coils and one of the
E−W coils together, we can also obtain NE, SE, SW and NW.

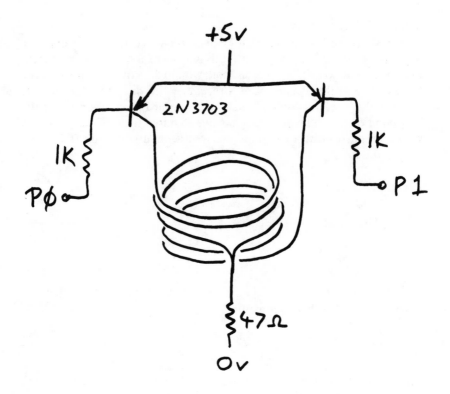

Figure 6.3   Bi-directional drive

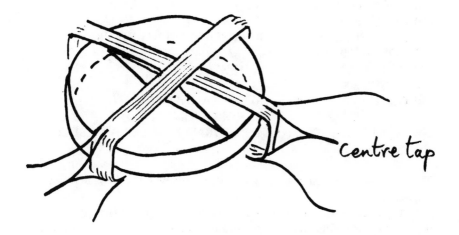

Figure 6.4   Two coils

Speed and acceleration control

Enter and run the following program:

```
10  DD = 56579:PO = 56577  :REM **** IF CBM 64
or
10  DD = 59459:PO = 59471  :REM **** IF PET

20  POKE DD,15             :REM BITS 0-3 ARE OUTPUTS
30  POKE PO,255-5          :REM BOTH COILS ON TO GIVE NE
40  GOSUB 200              :REM 1 SECOND DELAY
50  POKE PO,255-9          :REM NOW COILS GIVE NW
60  GOSUB 200
70  POKE PO,255-10         :REM NOW COILS GIVE SW
80  GOSUB 200
90  POKE PO,255-6          :REM NW
100 GOSUB 200
110 GOTO 30                :ROUND AGAIN FOR ANOTHER
                            REVOLUTION
200 FOR I = 1 TO 1000: NEXT:RETURN : REM DELAY 1 SECOND
    OR SO
```

The compass needle should now rotate, if somewhat jerkily, acting as a stepper motor.

Now you can try a variety of numbers in line 200 to set the speed of the motor. You will find that, if you aim too high, the motor will not even start. Try accelerating steadily by making the following changes:

Add line	5	V = 2000
Change line	200	FOR I = 1 TO V:NEXT
Add line	210	V = V-1
Add line	220	IF V< 50 THEN V = 50
Add line	230	RETURN

Now the delay will reduce progressively, until the top speed is reached. Try various values in line 220.

The speed will climb very slowly, rushing at the end. A steadier speed-up can be obtained with:

210 V = V*.995

You are now experimenting with techniques which you will need when you graduate to a genuine stepper motor. Of course, the program is still grossly inelegant, and is not exactly versatile. Nevertheless, the compass motor will already have taught you some of the pitfalls to look for:

1. Without drive, the motor does not retain its position.

2. Settling to a new position takes the form of a poorly-damped oscillation. At certain stepping speeds, there is a resonance so that the oscillations build up; the motor then stalls.

3. Movement at low speeds is 'lumpy'. This can be improved somewhat by doubling up on the applied steps, so that the sequence is N, NE, E, SE, S, SW, W, NW and back to N.

4. Sudden changes of speed will stall the motor.

5. There is no absolute position reference. Everything depends on the motor keeping in step.

Stepper software with some structure

Now let us try to introduce some 'style' into a new program, so that it will be of more general use. The codes which determine the coil polarities are best held in an array. I have a personal preference for putting all initialisation data at the end of the program, so that it does not obscure listings of the functional part. Thus the program will start with GOTO 10000, and all definitions will start at line 10000.

```
10000  DD = 56579:PO = 56577: REM **** CBM 64
```
or
```
10000  DD = 59459:PO = 59471: REM **** PET
```

```
10010  DIM DR(7):FOR I = 0 TO 7: REM DRIVE CODES
10020  READ J: DR(I) = 255 − J: NEXT I
10030  DATA 1,5,4,6,2,10,8,9:REM COILS IN ORDER N-S-E-W
10040  POKE DD,15: REM MAKE BITS 0−3 OUTPUTS
10050  GOTO 100 : REM HOUSEKEEPING DONE
```

The number of steps to move is held in variable DI (DIstance), whilst the direction is held in RO (ROtation) as a value ± 1. The current position is held in HEre, and the SPeed is set by variable SP. Now an appropriate section of program to command the movement could be:

```
 200  GOSUB 5000 : REM MOVE DISTANCE, ROTATION, SPEED
```

where the subroutine has been defined as:

```
5000  IF (DI< 1) OR (SP< 1) THEN RETURN
5010  FOR I = 1 TO DI
```

```
5020  HE = HE + RO :REM ADD ROTATION TO HERE
5030  POKE PO,DR(HE AND 7) :REM OUTPUT CODE TO PORT
5040  FOR J = 1 TO 1000 STEP SP:NEXT J
5050  NEXT I
5060  RETURN
```

The variable delay of line 5040 might appear a clumsy way to set the speed, but it is effective unless the value of SP is excessive. A more elegant technique, that of the 'binary-rate-multiplier', is described in the next chapter. It is useful for coordinating the movements of several steppers, but, because of an uneven stepping rate, the top speed is reduced.

To complete this program, you can add:

```
 10  GOTO 10000
100  PRINT "DISTANCE, ROTATION DIRECTION, SPEED"
110  INPUT DI,RO,SP
200  GOSUB 5000
210  GOTO 100
```

and you have a demonstration program enabling you to command a move from the keyboard. You should then be able to write a more elaborate program which builds an array of programmed moves and then executes them.

A second stepper motor can be added, driven from bits 4 to 7. This will enable you to make a plotter or a turtle, but will be a bit restrictive for a robot. You will need to use some clever.addressing techniques if up to eight motors are to be commanded from a single user port.

Power supplies

Before going into any more software detail, let us consider the electronic problems of interfacing one or more genuine stepper motors to the computer. The principles remain the same, but we must now be able to supply much greater currents. These are beyond the permitted drain which can be taken from the micro, and so a separate supply must be provided. You should be able to buy a 1 amp supply, variable from 4 to 10 volts, for under £30.00. Even so, this is scarcely enough current — although an overload will merely 'fold back' the output. The best answer may be to build the unstabilized supply described in Chapter 1, which will give around three amps output at + 7V and − 7V. Many stepper motors will require 12V or more to give of their best, and the power supply can be connected to give a single 14V supply — just by using the − 7V terminal as the negative connection and ignoring the centre terminal.

A lazy but risky alternative is to take your life in your hands and use a motor-car battery charger. This will probably give you up to four amps, but will need a large external capacitor — 10,000 microfarads or so. It will also give poor regulation, and will put nothing better than a four-amp fuse between your circuitry's well-being or annihilation. Still, it's better than buying an endless supply of batteries, unless you can afford rechargeables.

Interfacing hardware

The simple transistor will hardly have enough 'beta' to drive a stepper motor from the earlier circuit. However, you can buy Darlington transistors with much higher gain. They are, in fact, a pair of transistors in cascade, but have the disadvantage of a higher 'bottoming' voltage — they are less efficient in low voltage circuits. Nowadays, it is much more economic to buy multi-function chips than to buy individual transistors, and the RS 307−109 chip contains seven Darlingtons, complete with input resistors and protection diodes, for well under £2.00. Murphy's Law gets you, of course, because to drive two stepper motors you need eight outputs, not seven.

Another complication is that these circuits are sinks, not sources. The common point of the motor windings must therefore be connected to the positive supply, and the winding will be energised when the user port output bit is high, not low. The line of the computer program setting up the output patterns will have to be changed, leaving out the '255−' inversion, to become

10010 READ J:DRIVE(I) = J:NEXT I

Moreover, the program will have to poke PO and DD with 255 as early as possible, so that zeros will be output and the motor will not be incinerated under double its fair share of active currents.

With the change to the program described above, and with the circuit shown below, you should be confident of your ability to drive stepping motors and should be ready to build a simple turtle.

Multi-pole steppers

Stepper motors can have as many as 200 steps per revolution. As the inputs are switched through one 'electrical revolution', the motor only rotates through a few degrees. How is this achieved? **Figure 6.6** shows a rotor which, unlike a simple compass needle, has two North and two South poles. The motor coils are no longer wound directly across the rotor, but are wound on pairs of salient poles. When winding 1 is driven in the positive direction, let us say that poles 'A' become South and poles 'a' become

65

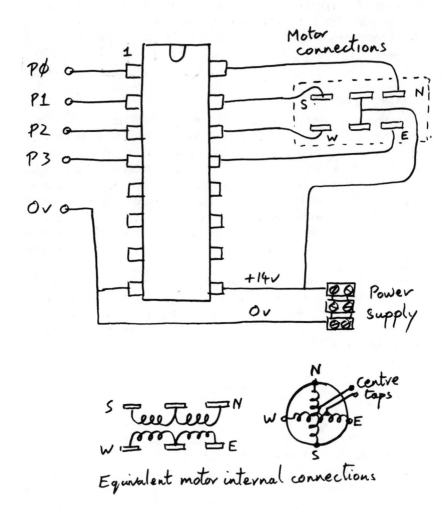

Figure 6.5 Darlington driver chip

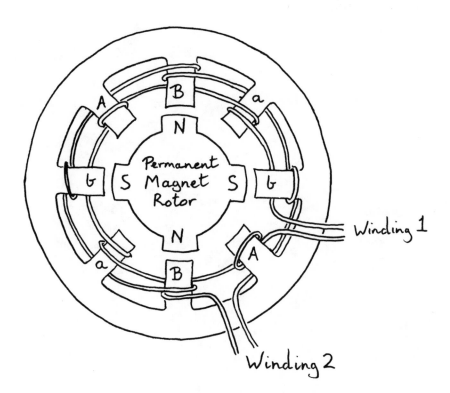

Figure 6.6   A four-pole stepper

North. When instead winding 2 is driven in a positive direction, poles 'B' become South and the rotor is pulled round through 45 degrees. After the windings have been stepped through one electrical revolution, winding 1 will again be driven positively, and the rotor will have made just half a turn. Put more poles on the rotor, and the ratio between electrical steps and rotation angle will increase.

CHAPTER 7
A Simple Turtle

If you come across an inverted soup bowl, wandering about and perhaps drawing shapes on a large sheet of paper, you have met a turtle. There is no attempt here to go into the intricacies of turtle graphics; instead the principles of the turtle serve as a good excuse for putting a pair of stepper motors to work.

Turtle fundamentals

The turtle is a simple 'wheelchair' system, propelled by two independent wheels on a diameter. Ball bearings or skids limit the resultant fore-and-aft toppling. To move straight ahead, both wheels rotate in step. To turn on the spot, one wheel rotates forwards while the other rotates backwards at exactly the same rate. If one wheel turns at exactly twice the speed of the other, the turtle will follow a circle with its centre one wheel-space from the slower wheel. Accurate movement calls for the motors being driven accurately in step — just the job for stepper motors!

At the centre of a 'genuine' turtle is a retractable pen, so that its perambulations can be used to draw shapes, or even graphs and illustrations. Let us think about that problem later.

Two stepper motors can be driven, with little complication, from the eight bits of the user port. With the aid of two multi-Darlington chips plus the experience of the last chapter, the task of making the motors rotate should give little trouble. The more difficult part is to make the software 'meaningful', so that a command structure can be based on the desired movements of the turtle without going into the gory details of the number of motor steps required for each gyration. Taking a 'top-down' look at the problem, we want to be able to type 'advance 100' to move 100 mm forwards, or perhaps 'turn, clockwise, 90'. Circles would be nice to add, with perhaps 'circle, clockwise, 200, 90' giving 90 degrees of a 200 mm radius circle. It might not even be over the top to add Cornu spirals to blend one radius with another — but not just at the moment. With graphics in mind, the further commands 'pen, up' and 'pen, down' complete the set. The task of working out where the turtle would wind up after a given manoeuvre can be performed on the command sequence by another subroutine, if required, although mechanical tolerances mean that the result will not be particularly accurate after a lengthy perambulation.

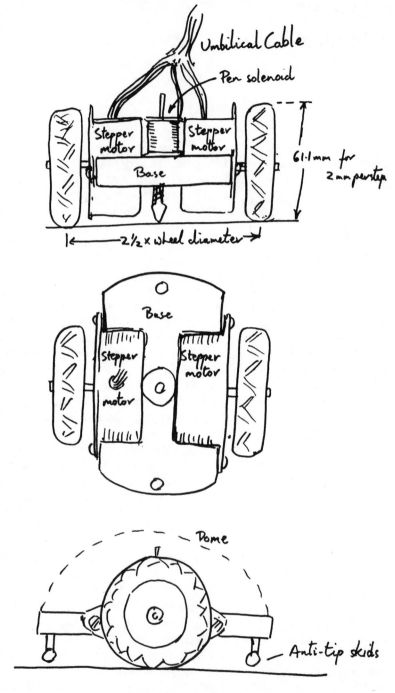

Figure 7.1   Views of a turtle

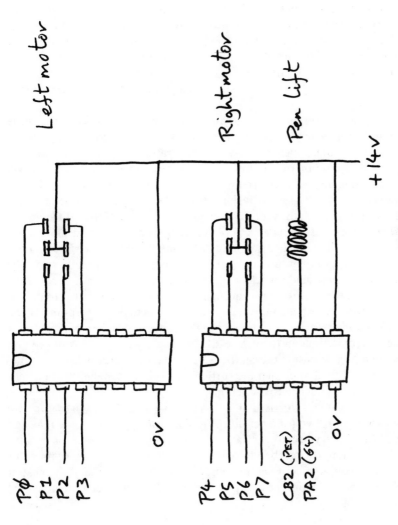

Figure 7.2   Circuit for turtle-darlingtons

71

Mechanical design

The suggested stepper motors are type ID35, made by Philips and distributed by Impex of Richmond at a price around £12.00. They have 48 steps per rev., ie 12 electrical revolutions per mechanical revolution. If you drive the motor in half-steps, ie N, NE, E, SE, S, SW, W, NW, you will now have 96 half-steps per revolution of the wheel. Suppose that your wheels are 80 mm in diameter, then they will have a circumference of around 250 mm and each half-step will give a movement of about 2.6 mm. If you don't mind making or trimming your own wheels, then a diameter of 2*96/pi = 61.1 mm will give exactly 2 mm per half step — but you would be best to buy the nearest larger size from the model-shop, and accept a slightly odd scale factor.

When the motors are driven equally in opposite directions, the turtle rotates about its centre. If each wheel makes one revolution, then the turtle will turn through (diameter-of-wheel/separation-of-wheels) revolutions. Make the separation two-and-a-half times the diameter, and each step will give just one degree. If the motors are driven at unequal speeds, the distance advances will be given by the average of the (signed) number of steps, whilst the turtle will turn through an angle equal to half their difference.

The chassis can be made from plywood or even balsa wood, since it has very little work to do. The skids can be formed from lightweight cupboard ball-catches, although a couple of bent paper clips will really serve the purpose. They should just clear the ground, so that only one touches the ground at a time. Most of the mechanical load will be due to the umbilical cable, and this must be connected to the turtle at a high central point. If you sacrifice some sort of plastic bowl to make a cover, then the cable can safely emerge from a hole in the centre. If, however, your turtle is naked you should mount a mast in the centre — not too tall, or the turtle will topple. The cable should approach the turtle from above, dangling from a supporting string attached to the ceiling.

At first sight you will need at least a dozen conductors in the cable, five for each motor, two for a pen-lift plus more for any sensors you may add later. At a pinch you can get away with two less, sharing a common positive power line, but this may be a false economy since the resistance of the cable can cause coupling between the motor drives. Ribbon cable is the neatest solution, but far from the cheapest. Perhaps I should recommend a good book on plaiting.

Control strategies

We can make up an algorithm for converting the commands into demanded motor half-steps (from now on, let us call them just 'steps') as follows. Let us assume a wheel diameter of 61 mm and a separation of 2.5*61 = 152.5 mm. See **Table 7.1**.

Table 7.1

COMMAND	LEFT MOTOR STEPS	RIGHT MOTOR STEPS
Advance	distance/2	distance/2
Turn, cw	+ angle	− angle
Turn, acw	− angle	+ angle
Circle, cw	angle*(radius/115 + 1)	angle*(radius/115 − 1)
Circle, acw	angle*(radius/115 − 1)	angle*(radius/115 + 1)

Now the command interpreter must 'talk to' a motor control module, which will accept commands in the form of the number of steps each motor must move. An extra command, 'speed, 20', can adjust a general variable which need not feature in the syntax. Let us use a subroutine to command the motors, and let us define this at line 8000 onwards.

Varying the speed of a single motor can be done with a simple variable delay, but to drive two motors at different speeds calls for a different concept, the binary-rate-multiplier. Suppose that the left motor must move 100 steps, whilst the right motor must move only 67. Then we first construct the ratio of the two, in this case 0.67. Each time round the loop we step the left motor, but the right motor may or may not need to step. To make the decision we keep adding the ratio to another variable, say T. If T is now greater than 1, the motor is stepped and T is reduced by 1. Sounds confusing? Then let's try an example, in **Table 7.2**.

Table 7.2

LEFT MOTOR POSITION		T	RIGHT MOTOR POSITION		
	0	0		0	
Step	1	.67		0	
Step	2	1.34	Step	1	T *becomes* 0.34
Step	3	1.01	Step	2	T *becomes* 0.01
Step	4	.68	Step	2	
Step	5	1.35	Step	3	T *becomes* 0.35
. . . . .					
. . . . .					
Step	97	.99	Step	64	
Step	98	1.66	Step	65	T *becomes* 0.66
Step	99	1.33	Step	66	T *becomes* 0.33
Step	100	1.00	Step	67	T *becomes* 0.00

So we arrive at the end of the movement with each motor having taken the correct number of steps. This is the principle behind most graph-plotting routines for drawing oblique straight lines. The result is slightly improved if T starts with the value 0.5, since this causes the unevenness to be shared out symmetrically along the line. In the example above, the only occurrence of three right-motor steps in a row is at the end of the movement; had T started with the value 0.5 they would have occurred in the middle.

Motor control software

We can now define the subroutine to MOVE the motors, where the number of steps for the left motor is stored in LM and for the right motor in RM.

```
8000  AL = ABS(LM):AR = ABS(RM):REM ABSOLUTE VALUES OF
      STEPS
8010  SL = SGN(LM):SR = SGN(RM):REM AND SIGNS OF
      DIRECTIONS
8030  IF AR + AL = 0 THEN RETURN:REM NO MOVE, GO HOME
8040  IF AR> AL THEN 8200:REM DEAL WITH THIS
      SEPARATELY

8100  RA = AR/AL: T = 0.5:REM RATIO OF MOVES
8110  FOR M = 1 TO AL:REM HERE WE GO
8120  GOSUB 9000:REM STEP LEFT MOTOR DIRECTION SL
8130  T = T + RA
8140  IF T> 1 THEN GOSUB 9100:T = T − 1: REM RIGHT MOTOR
8150  GOSUB 9500:NEXT M: RETURN:REM THAT'S ALL, GO
      HOME

8200  RA = AL/AR: T = 0.5:REM RIGHT MOVE > LEFT MOVE
8210  FOR M = 1 TO AR
8220  GOSUB 9100:REM RIGHT MOTOR EVERY TIME
8230  T = T + RA
8240  IF T> 1 THEN GOSUB 9000:T = T − 1: REM LEFT MOTOR
8250  GOSUB 9500:NEXT M:RETURN
```

This still leaves us with the 'bottom-up' task of writing the motor drivers. We start with housekeeping at line 10000:

```
10000  LP = 0:RP = 0:SP = 100:REM MOTOR POSITIONS,SPEED

10010  PO = 56577:DD = 56579: REM PORT, DATA DIR *** CBM 64
```
or
```
10010  PO = 59471:DD = 59459: REM                    *** PET

10020  DIM LD(7),RD(7):      REM TWO ARRAYS FOR MOTOR
                             DRIVES
10030  FOR M = 0 TO 7
10040  READ J: LD(M) = J: RD(M) = 16*J: NEXT
```

```
10050  DATA 1,5,4,6,2,10,8,9:REM COILS IN ORDER N-S-E-W
10060  POKE DD,255:POKE PO,0:REM MAKE OUTPUTS, SET TO
       ZERO
10070  GOTO 100
```

Then we add the motor drivers, and a delay which depends on SPeed:

```
9000  LP = (LP + SL) AND 7: REM LEFT MOTOR NEW POSITION
9010  POKE PO, (PEEK(PO) AND 240) + LD(LP)
9020  REM MIX NEW LEFT MOTOR DRIVE WITH OLD RIGHT,
      OUTPUT
9030  RETURN
9100  RP = (RP + SR) AND 7: REM RIGHT MOTOR NEW POSITION
9110  POKE PO, (PEEK(PO) AND 15) + RD(RP)
9120  RETURN
9500  FOR D = 1 TO 1000 STEP SP:NEXT
9510  RETURN
```

(Using a modicum of cunning, you should be able to rewrite the motor procedures into a single procedure with two arguments QL and QR present to SL, SR or zero. You should then be able to tidy up the move procedure to make it less lumpy. The inelegant procedures here are designed to be easier to understand.)

Before adding the clever stuff, trouble-shoot these modules with a 'jiffy program':

```
 10  GOTO 10000
100  PRINT "LEFT MOTOR, RIGHT MOTOR"
110  INPUT LM,RM
120  GOSUB 8000
130  GOTO 100
```

and if the result does not look too good, try something even simpler:

```
100  SPEED = 10:SL = 1
110  GOSUB 9000:GOSUB 9500: GOTO 110
```

to get down to bedrock. If all else fails, take manual control by POKEing PO to various values, and get out your trusty test meter.

Pen lift

Whilst we are dealing with the 'nuts and bolts', let us have a look at the pen lift. Having thoroughly used up the bits of P0 to P7, the only convenient

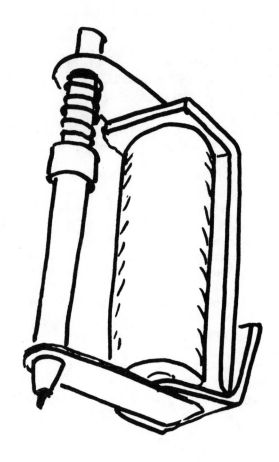

Figure 7.3   Pen lift from post office relay

user port bit left is now pin M, position 11 of the connector strip, which is PA2 (64) or CB2 (PET). Without wishing to tangle too closely with the intricacies of the peripheral control register of the VIA, it is safe to reveal that POKE DD + 9,14*16 (POKE $E84C,$E0) will set the PET's CB2 high, while POKE DD + 9,12*16 will set it low. Now pin M can be wired to another channel of a Darlington chip (using up one of the six spare channels) to give a signal beefy enough to drive a solenoid. For the 64 we can add:

```
7000  D = PEEK(PO−1) AND 251: REM PORT A, BIT 2 LOW:
                         *** CBM 64
7010  IF P = 0 THEN D = D + 4  : REM ENERGISE TO LIFT PEN
7020  POKE PO−1,D:RETURN:REM OUTPUT TO PORT A
```

or for the PET we can instead add:

```
7000  IF P = 0 THEN POKE DD + 9,224:RETURN:   REM ENERGISE
                                              TO LIFT PEN
7010  POKE DD + 9,192:RETURN:   REM DEENERGISE, DROP
                                PEN
```

This does not answer your problem of finding a pen-lift solenoid to drive. A commercial solenoid can easily be bought, but is likely to be heavy and over-powered. It does not take much to lift a ball-point, and even less to lift a felt-tipped pen, and you can substitute a little dexterity for a lot of power consumption. An old post-office relay can, with the removal of the contact assembly, provide more than enough lift. You may need to fiddle a little with the pen height, but, provided your wheels are not eccentric, you should get acceptable results.

Simple command interpreter

Now we are ready for the command interpreter. This could be written most elegantly and almost incomprehensively with searches in command lists. Instead let us try a 'knife and fork' job, which will simply perform each command immediately. It can be adapted later to memorise and edit a command sequence. The task of inputting the commands is not made easier by the motley assortment of arguments they can take. We have on the one hand 'ADVANCE, 200', and on the other 'CIRCLE, CW, 45, 150', so the user will welcome some user-friendly guidance. Let us put the parsing routine at 1000:

```
100  GOTO 1000
```

```
1000 PRINT"COMMAND:   ";:INPUT A$
1010 IF A$< > "ADVANCE" THEN 1100
1020 PRINT"DISTANCE:   ";:INPUT DS
1030 LM = DS/2:RM = DS/2
1040 GOSUB 8000:GOTO 1000

1100 IF A$< > "TURN" AND A$< > "CIRCLE" THEN 1300
1110 PRINT"CW/ACW:   ";:INPUT B$
1120 PRINT"ANGLE (DEG):   ";:INPUT AG
1130 IF B$ = "ACW" THEN AG = -AG
1140 IF A$ = "TURN" THEN LM = AG:RM = -AG:GOSUB
     8000:GOTO 1000
```

Turtle Program

```
10 GOTO 10000
1000 PRINT"COMMAND: ";:INPUT A$
1010 IF A$<>"ADVANCE" THEN 1100
1020 PRINT"DISTANCE: ";INPUT DS
1030 LM=DS/2:RM=DS/2
1040 GOSUB 8000:GOTO1000
1100 IF A$<>"TURN" AND A$<>"CIRCLE" THEN 1300
1110 PRINT "CW/ACW: ";:INPUT B$
1120 PRINT "ANGLE: ";:INPUT AG
1130 IF B$=".ACW" THEN AG=-AG
1140 IF A$<>"TURN" THEN 1200
1150 LM=AG:RM=-AG:GOSUB 8000:GOTO 1000
1200 PRINT "RADIUS: ";:INPUT R
1210 IF B$="ACW" THEN R=-R
1220 LM=AG*(R/115+1):RM=AG*(R/115-1)
1230 GOSUB 8000:GOTO1000
1300 IF A$<>"SPEED" THEN 1400
1310 PRINT "SPEED WAS ";SP
1320 PRINT "NEW SPEED: ";:INPUT SP:GOTO 1000
1400 IF A$<>"PEN" THEN 1500
1410 PRINT "UP/DOWN: ";:INPUT B$
1420 P=1:IF B$="UP" THEN P=0
1430 GOSUB 7000:GOTO 1000
1500 PRINT "SORRY - CAN'T RECOGNISE COMMAND"
1510 PRINT "ADVANCE, TURN, CIRCLE, SPEED, PEN"
1520 GOTO 1000:REM ADD NEW COMMANDS AT 1500
7000 D=PEEK(PO-1) AND 251: REM *** CBM 64
7010 IF P=0 THEN D=D+4:REM LIFT PEN
7020 POKE PO-1,D: RETURN
8000 AL=ABS(LM): AR=ABS(RM)
8010 SL=SGN(LM): SR=SGN(RM)
8030 IF AR+AL=0 THEN RETURN
8040 IF AR>AL THEN 8200
```

```
8100 RA=AR/AL: T=.5
8110 FOR M=1 TO AL:GOSUB 9000:T=T+RA
8140 IF T>1 THEN GOSUB 9100: T=T-1
8150 GOSUB 9500:NEXT M:RETURN
8200 RA=AL/AR: T=.5
8210 FOR M=1 TO AR:GOSUB 9100:T=T+RA
8240 IF T>1 THEN GOSUB 9000: T=T-1
8250 GOSUB 9500:NEXT M:RETURN
9000 LP=(LP+SL) AND 7
9010 POKE PO,(PEEK(PO) AND MR)+LD(LP)
9020 RETURN
9100 RP=(RP+SR) AND 7
9110 POKE PO,(PEEK(PO) AND ML)+RD(RP)
9120 RETURN
9500 FOR D=1 TO 1000 STEP SP:NEXT:RETURN
10000 LP=0:RP=0:MR=240:ML=15:      REM MASKS
10010 PO=56577: DD=56579:      REM *** CBM 64
10020 SP=100:DIM LD(7),RD(8)
10030 FOR M=0 TO 7
10040 READ J: LD(M)=J: RD(M)=16*J: NEXT
10050 DATA 1,5,4,6,2,10,8,9:      REM N-S-E-W
10060 POKE DD,255: POKE PO,0
10070 GOTO 1000
```

CHAPTER 8
Interfacing a Robot

A fully-fledged robot has seven degrees of freedom, that is to say it requires seven independent motors to drive it. The 'end effector' (a fancy term for 'hand') must be able to move in three dimensions, and for any given position it should be able to swivel about three more axes. A further channel is needed for 'open' or 'close', although this is often a simple on/off valve working a pneumatic gripper. Educational robots, such as the Armdroid, sacrifice one of the 'wrist' axes, but give continuous grip movement. This reduces the number of channels to six. If these are driven by stepper motors, how can we interface them to the computer? In the last chapter, two channels of stepper motor were interfaced to the user port with the use of all eight bits, so for six channels we must find some new technique to command them all. It is necessary to include an address as part of the user-port data, which can be decoded within the robot itself.

Multi-stepper control

The user port provides eight bits: a stepper motor needs four bits to define a half-step position (unless you are happy interfacing using the scale of three: N, off, S). That leaves four bits for housekeeping. From three of these bits, an address can be constructed to address eight channels. The addressed channel will now capture the motor signals in a four-bit latch, and carry on driving the motor lines until told to do otherwise. Now we need a 'strobe' signal as well, so that we can tell the circuitry 'the motor lines have finished changing, the address lines are settled, catch this data now and use it'.

The connections to the somewhat ageing Armdroid on which the programs of this chapter have been tried out are as follows:

PB0	PB1	PB2	PB3	PB4	PB5	PB6	PB7
Strobe	------Channel------			N---------E---------W---------S			

Note that the 'compass' bits are shuffled in comparison with the last chapter, and so the data statements for the codes are different. Putting the 'channel' bits in a shifted position does complicate the code calculation a little, but everything comes out in the wash.

Since the strobe will be active when low, the procedure for outputting a new command is as follows:

1. Look up the code for the desired motor position.
2. Add on 2*(channel number), hold the result in CO, say.
3. Set the strobe bit (bit 0) high in CO.
4. Output CO to the user port.
5. Set bit 0 of CO low; output CO to the user port.
6. Set bit 0 of CO high again; output CO to the user port.

Putting the algorithm into software

Let us adopt our usual technique of defining data and arrays 'up in the sky', with housekeeping at 10000:

```
10000 DD = 59459:PO = 59471:KB = 547:KO = 255:REM ANTIQUE
      PET ******
10000 DD = 59459:PO = 59471:KB = 166:KO = 255:REM 8000, 4000
      PET ****
10000 DD = 56579:PO = 56577:KB = 197:KO = 64 :REM
      COMMODORE 64 *****
10010 DIM DR(7): REM DRIVE VALUES WITH STROBE ALREADY
      HIGH
10020 FOR I = 0 TO 7: READ J: DR(I) = 16*J + 1: NEXT
10030 DATA 1,3,2,6,4,12,8,9: REM N-E-S-W
10040 MA = 254: POKE DD,255 :REM SET TO OUTPUTS
```

Now we can look at a subroutine for driving one of the motors. If the channel number is set in variable CH, whilst the step angle is held as value V (measured as number of steps from the switch-on position), then the following routine will output the required code to the motor:

```
9000 CO = DR(V AND 7) + CH*2
9010 POKE PO,CO: REM OUTPUT THE CODE, STROBE BIT HIGH
9020 POKE PO,CO AND MA :REM 'AND' WITH MASK — STROBE
     LOW
9030 POKE PO,CO: REM STROBE HIGH AGAIN
9040 RETURN
```

This routine takes one or two short cuts from the algorithm above, and will simply set up one motor drive to command a given position.

Troubleshooting the connections

For the Armdroid's end of the connections, you will have to refer to the handbook — the edge-connector has been modified in recent issues. If you have constructed your own circuit (perhaps from the one given later in this chapter) then it should already be familiar to you.

Apart from reassuring yourself that something will really happen, the next test will establish which channel numbers control which axes of the robot, and in which direction. Enter the following simple test program. Since it will get progressively overwritten as you work through the chapter, you might like to save it at each stage for future use.

```
   10  GOTO 10000
  100  INPUT "CHANNEL NUMBER, DISTANCE":CH,DI
  110  FOR V = 0 TO DI STEP SGN(DI)
  120  GOSUB 9000:NEXT
  130  GOTO 100
10900  GOTO 100:REM AT END OF HOUSEKEEPING
```

Now check out each of the channels in turn, entering values from 0 to 7 for the channel number, and around 50 or − 50 for the distance. Two of the eight possibilities will of course have no effect, since only six channels are used. Make a careful note of the axis and the direction, in terms of up, down, pivot left, right, forwards, backwards, gripper open and close. The Armdroid uses two motors at a time to rotate the wrist, and to swivel the wrist up and down. Make a note of which does which. A program with a machine-code output routine can afford the time to unscramble separate commands to twist and tilt the wrist. In BASIC, this makes the system rather slow, so here we will at first drive just one motor at a time.

Keypress commands

Now let us add a routine so that holding down a key will drive a motor. It is important to display a menu of keys on the screen, showing which key does what — nothing is more frustrating than having to guess. To get smart-looking upper and lower cases legends, put the machine into lower-case mode before typing in the program. Of course, when listed in this mode all the commands in the software will appear in lower case, but it is clearer to show them here in the usual capitals.

Now we must allocate a key to each movement, up, down, left, right, forwards, back, open and close, plus wrist up and down and rotate. The keys U, D, L, R, F, B, O and C are obvious enough as choices, but we must choose four more for the wrist — how about W, Q, T and Y? To each key will correspond a channel number and a direction. We must add a set of data statements to the housekeeping to sort them out — use your results from above to correct the values below, to give the correct movements. The values given here are the ones which Richard has sorted out for his particular Armdroid:

```
10100  DIM CH(5),HE(5):FOR I = 0 TO 5:READ J
10110  CH(I) = 2*J:NEXT:REM CHANNEL CODES
10120  DATA 1,3,5,4,2,6:REM U/D, L/R, F/B, O/C, WRIST
10130  C$ = "UDLRFBOCWQTY":REM COMMAND KEYS
```

You should swap the pairs of letters in C$ as necessary so that the first corresponds to a positive direction, the second negative. The array HEre(I) is used to remember the current position of each motor, and will become very useful later. Now we patch in a routine at 1000 to display the menu, read the keypress, interpret the command and execute it:

```
1000  PRINT CHR$(147);  "Up         Down"
1010  PRINT"Left        Right"
1020  PRINT"Forward     Back"
1030  PRINT"Open        Close"
1040  PRINT"Wrist —     Q"
1050  PRINT"Twist —     Y"
1100  GET A$:K = PEEK(KB):IF K = KO OR A$ = "" THEN 1100
1150  FOR I = 1 TO LEN(C$):IF A$< > MID$(C$,I,1) THEN
      NEXT:GOTO1000
1160  J = I:I = LEN(C$):NEXT: REM CLOSE LOOP NEATLY
1170  CH = INT((J − 1)/2):DI = 2*(J AND 1) − 1:V = HE(CH)
1180  V = V + DI:GOSUB9000:REM TAKE A STEP
1190  IF PEEK(KB) = K THEN 1180:REM KEEP STEPPING IF
      PRESSED
1200  HE(CH) = V:GOTO1000
```

To make this work, we need to change line 9000 to make use of our channel unscrambler:

```
9000  CO = DR(V AND 7) + CH(CH)
```

We must also remove the old 100−130, writing instead:

```
100  PRINT CHR$(14):GOTO 1000
```

As it now stands, the program should run reasonable quickly. There is, however, a very simple dodge to speed it up a little. To access a variable, BASIC must scan down the list of variables in the order that they were first encountered until it comes to the correct one. Thus the variables declared earliest will operate the fastest. Looking through the program, we see that CO, PO, V, DI, CH, KB and K are used every step, some several times. If we change line 10 to the strange-looking form:

10 DIM CO,PO,V,DI,CH,I,R,KB,K:GOTO 10000

then there should be a satisfying increase in speed. (I and R have been included for later use.)

Programmed movements

When every move must be controlled from the keyboard, the robot is just a toy. If, however, we can program into it a sequence of movements which it can perform automatically, then we can start to explore its serious use. We therefore need to be able to record each target point, and we need a second command level which will let us switch between the 'teach-mode' section (the program we have already tested) and the routines which will drive the robot automatically. For this, we use a second menu starting at line 100:

```
100  PRINT CHR$(147);"Teach,          Perform,"
110  PRINT "Repeat,      Clear,"
120  PRINT "Save,        Input"
130  GET A$: IF A$ = ""THEN 130: REM WAIT FOR A KEY-PRESS
140  FOR I = 1 TO 6:IF A$< > MID$("TPRCSI",I,1) THEN
     NEXT:GOTO100
150  J = I:I = 6:NEXT: REM CLOSE THE FOR...NEXT LOOP
     NEATLY
160  ON J GOTO 1000,5000,5100,2000,6000,7000
```

So far, we have only written the 'teach' routine at 1000, and even this needs some modification. We must allow the selection of any point to be added to the manoeuvre, and we must allow a return to command mode. We therefore fill in the gaps in the program with:

```
1060  PRINT"Point          End teach"
1110  IF A$ = "E" THEN 100
1120  IF A$< > "P" THEN 1150
1130  IF NP = MP THEN 1000:REM TOO MANY POINTS
1140  NP = NP + 1:FOR  I = 0  TO  5:PT(I,NP) = HE(I):NEXT:GOTO
      1000
```

Now for some more housekeeping. The array PT(5,MP) must be declared to hold the points. The limit, MP, can conveniently be set at 20, but if you wish you can choose a much bigger number.

10300 NP = 0:MP = 20:DIM PT(5,MP)

To keep track of progress, we can display the present coordinates and the number of points set with:

1080 PRINT:PRINT NP;"POINTS"
1090 FOR I = 0 TO 5:PRINT HE(I):NEXT

That completes the teach mode section of the program. We are left with the task of writing the routine to perform the set of actions.

Routines to perform a manoeuvre

Having remembered the moves, we can perform them by looking up target points in turn, and then calling a routine at 8000 to move from HEre(I) to TArget(I). If we wish to perform them just once, then we can GOTO 5000, and:

```
5000  IF NP = 0 THEN GOTO 100:REM NO POINTS
5010  FOR P = 1 TO NP
5020  FOR I = 0 TO 5:TA(I) = PT(P,I):NEXT
5030  GOSUB 8000
5040  NEXT P
5050  GOTO 100
```

If we wish to repeat the cycle until a key is pressed, then we can use a section of program at 5100:

```
5100  IF NP = 0 THEN GOTO 100
5110  FOR P = 1 TO NP
5120  FOR I = 0 TO 5
5130  TA(I) = PT(P,I):NEXT
5140  GOSUB 8000
5150  NEXT P
5160  GET A$:IFA$ = ""THEN 5110: REM NO KEY, ROUND AGAIN
5170  GOTO 100
```

This has still not resolved the problem of what to put at 8000. For best speed of response, we will at first move just one axis at a time, although we will go on to consider diagonal movements. After declaring the array TArget(5) with:

10200 DIM TA(5)

we can compare each channel of TArget against HEre, and, if necessary, move accordingly.

```
8000  FOR CH = 0 TO 5
8010  IF TA(CH) = HE(CH)THEN NEXT:RETURN
8020  FOR V = HE(CH) TO TA(CH) STEP SGN(TA(CH)-HE(CH))
8030  GOSUB 9000:NEXT:HE(CH) = TA(CH)
8040  NEXT:RETURN
```

Now, to complete the command routines we can add a 'clear' one-liner:

```
2000  NP = 0:GOTO100
```

and we should also add routines at 6000 and 7000 to record and retrieve a set of movements. For now, plus the lines with:

```
6000  GOTO 100
7000  GOTO 100
```

and create your own routines when you have checked out the rest of the program.

As it stands, the program will act as a quite acceptable robot driver, but the execution of the manoeuvres, one axis at a time, will seem inelegant. It would be far better to interpolate the movements with all axes firing together. Unfortunately, the program which follows in the next section is excruciatingly slow in its performance, and the only really satisfactory way to achieve the result is by using machine code.

Simultaneous movements

We want a routine which will set up all six channels of the robot, and will interpolate a manoeuvre so that all channels can be made to move at once. This is another task for the binary-rate-multiplier, this time firing on six cylinders. Our starting position is HEre, an array with six elements, one for each motor axis. The destination is held in TArget, and we can work through all six axes, finding which one demands the greatest change. Now it is a straightforward job to calculate the ratios, and to make a step according to the overflow of a REgister, just as in the last chapter. We need some more arrays along the way, and the housekeeping routine at 10200 becomes:

```
10200  DIM TA(5),RA(5),WA(5),RE(5)
10210  FOR CH = 0 TO 5:HE(CH) = 0:RE(CH) = .5
10220  V = 0:GOSUB9000:NEXT:REM ZERO MOTORS
```

Now we can write a subroutine to move all six channels from HEre to the TArget position:

```
8000 REM MOVE FROM HERE TO TARGET : FIRST FIND LONG-
     EST MOVE
8010 RM = 0:FOR CH = 0 TO 5
8020 RA(CH) = ABS(TA(CH)-HE(CH)):REM   WORK   OUT   DIS-
     TANCE AND
8030 WA(CH) = SGN(TA(CH)-HE(CH)):REM   DIRECTION   FOR
     EACH AXIS
8040 IF  RA(CH)> RM  THEN  RM = RA(CH):REM  FIND  MAX
     DISTANCE
8050 NEXT:IF RM = 0 THEN RETURN : REM NO MOVE
8060 FOR CH = 0 TO 5
8070 RA(CH) = RA(CH)/RM:NEXT:REM RATES NOW IN
     RANGE 0 TO 1

8100 FOR R = 1 TO RM: REM NOW WE ARE READY TO MOVE
8110 FOR CH = 0 TO 5
8120 RE(CH) = RE(CH) + RA(CH)
8130 IF RE(CH)< 1 THEN 8160
8140 RE(CH) = RE(CH)-1:HE(CH) = HE(CH) + WA(CH)
8150 V = HE(CH):GOSUB 9000 :REM MOVE MOTOR
8160 NEXT CH
8170 NEXT R
8180 RETURN : REM NOW HERE = TARGET
```

Lines 8100 to 8180 are written with the aim of being easy to understand. Some improvement in speed can be made at the expense of clarity. After trying the first version, try substituting this:

```
8100 FOR R = 1 TO RM:FOR CH = 0 TO 5:I = RA(CH):IF I = 0 THEN
     8130
8110 I = I + RE(CH):IF I< 1 THEN RE(CH) = I:NEXT:NEXT:
     RETURN
8120 RE(CH) = I-1:V = HE(CH) + WA(CH):HE(CH) = V:GOSUB 9000
8130 NEXT:NEXT:RETURN
```

8140−8180 are now redundant and should be deleted.

Even if you use only the program listed here, you will be able to teach the robot a routine which it will perform with interpolated movements. I still don't claim that the execution will be fast — you will soon want to perform the binary-rate-multiplier functions in machine code, and include a ramp

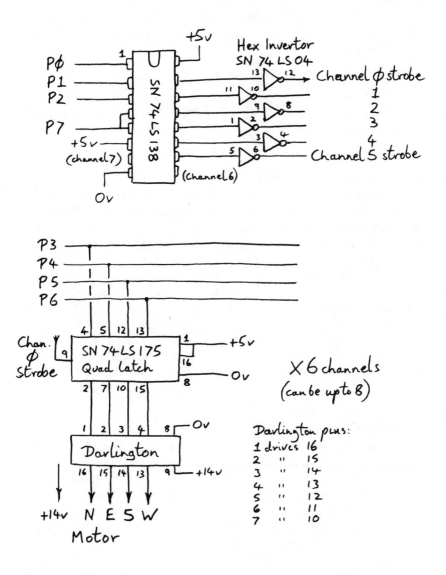

Figure 8.1   Circuit for 6-axis robot — decoder, latches, drivers

89

speed-up routine to achieve top speed without breaking away. Some elegant programs (eg MEMROB, written in collaboration with Tim Dadd and distributed by Colne for using an Armdroid with a PET) link the interpolation and output functions to the computer's interrupt, and communicate between BASIC and machine code by planting values in an array of variables. In this way, the BASIC part can do its 'thinking', working out the speed ratios for the next move at the same time as the present move is being made. The listing of MEMROB has baffled multitudes, and has little teaching value.

Building your own robot

You will first need half-a-dozen stepper motors. The ID35 mentioned in the last chapter will do nicely — it is the one used in the Armdroid. The Darlington drivers have been covered pretty thoroughly in Chapter 6. That leaves only the channel decoders and the four-bit latches — and the power supply. The supply detailed in Chapter 1 should be adequate, and will cost less than a single stepper motor. A circuit diagram for the 'innards' of the robot is drawn in **Figure 8.1**, and this should do all that you need.

For a do-it-yourself mechanical design, you can either follow close on the heels of the commercial robots, or you can be more adventurous. When you start to examine the number of ways you can link six motors together, the choice is amazing. Your geometry can be cartesian, polar, cylindrical-polar or a variety of strange hybrids. Let us start by looking at the 'conventional' robots.

Robot anatomy

The first axis of movement is a rotation of the whole assembly about the vertical. You can 'humanise' this by thinking of it as swivelling about the waist. Next, the shoulder joint allows the arm to tilt up and down, so that, using these two motors alone, the hand could reach any point on the surface of a sphere. Next comes the elbow joint. As this bends, the arm is effectively shortened, although the hand now moves in a way which needs more and more trigonometry to describe it. In principle, the robot should now be able to reach any point within a sphere, but if, for example, the upper arm is not the same length as the forearm, there will be some unreachable zones. The wrist joint should now be able to swivel both up-and-down and left-to-right. The Unimation Puma instead uses a movement like the human wrist, where the up-and-down hinge is an axis which can in turn swivel about the line of the forearm. The Armdroid leaves one of these movements out. Now the Puma can in effect line up a screwdriver with a screw in any position; the final axis twists the screwdriver to drive the screw.

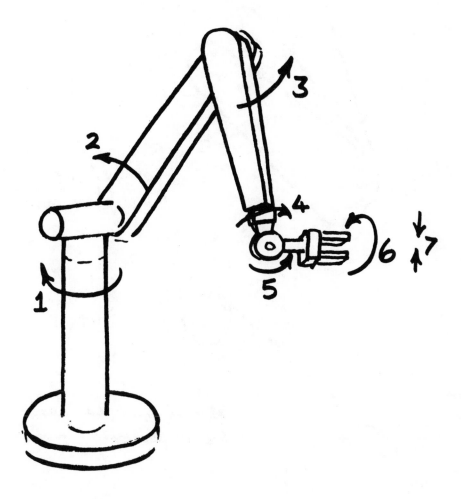

Figure 8.2   Axes of a 'conventional' robot

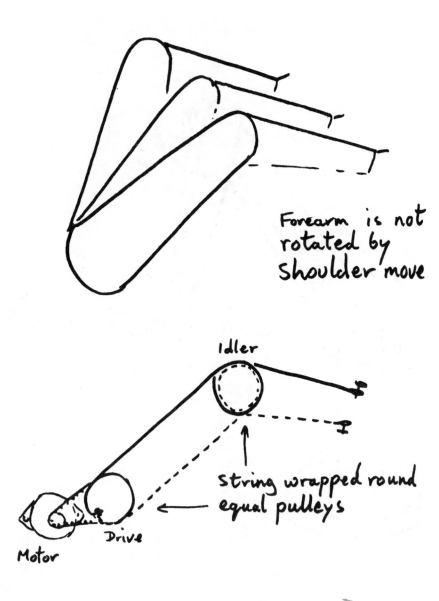

Forearm is not rotated by shoulder move

Idler

string wrapped round equal pulleys

Drive

Motor

Figure 8.3   Stringing to obtain parallel forearm movement

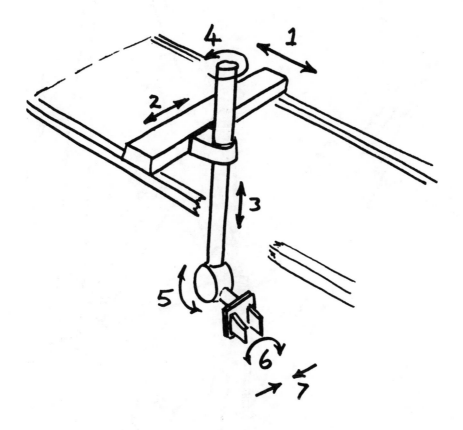

Figure 8.4   Cartesian robot arrangement

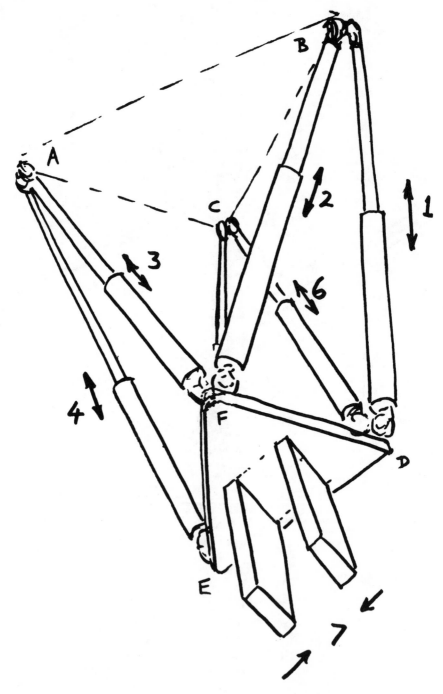

Figure 8.5   A variation on the 'Gadfly'

Some sophisticated robots, such as the Puma, perform laborious computations, allowing the user to specify that the hand should move in a straight line and that the tool should not rotate in space. New positions are calculated for the motor axes up to forty times per second, and the movement is then smoothed out by 'rate control', similar to the techniques described earlier. It is a challenging exercise to try!

Even when the geometry is settled, there are many ways to connect the motors to the axes. The Puma uses brute force, so that the entire leverage of the arm and its load will appear at the shoulder joint. The Armdroid on the other hand uses a cunning bit of string-work, so that as the shoulder rotates the upper arm, the forearm remains parallel to its former position. This effectively halves the leverage of the load on the shoulder motor — although it does not do a lot for the string which is annoyingly apt to break.

The IBM robot is cartesian, and bears a strong resemblance to an over-grown graph plotter for controlling the X and Y axes. The Z axis is an even more overgrown pen-lift, raising and lowering a bar which can rotate to provide the first of the wrist axes. All the problems of straight-line movement are solved at a stroke, but tracks and pulleys are now needed in place of pivots and levers.

When computing power is let loose, anything goes. Considerable industrial research is being put into developing a device using six extending rods, driven by motors and leadscrews. Imagine a triangle ABC fixed to the floor, with the tool attached to a movable triangular plate DEF. The plate is held up by six rods AE, AF, BF, BD, CD and CE. As these vary in length, so the plate can be moved in three dimensions and rotated about three axes. If you have an idle moment, calculate the relationship between the lengths and the position of the plate, or more especially the lengths required to place the plate in any particular position — a prize is offered for the simplest solution! No wonder it is called the 'Gadfly'.

These variations hardly scratch the surface of the possible combin-ations. If you connect up the six stepper motor channels, you can try any number of 'lash-ups' using cardboard, string and balsa wood before immortalising your design in aluminium or steel. Good luck!

Robot Control Program

NB. Change line 10000 if using a PET.

```
10 DIM CO,PO,V,DI,CH,I,R,KB,K: GOTO 10000
100 PRINT CHR$(147);"TEACH,     PERFORM,"
110 PRINT"REPEAT,    CLEAR"
120 PRINT"SAVE,      INPUT"
130 GET A$: IF A$="" THEN 130:REM AWAIT KEYPRESS
```

```
140 FOR I=1 TO 6
145 IF A$<>MID$("TPRCSI",I,1) THEN NEXT:GOTO 100
150 J=I:I=6:NEXT: REM CLOSE 'FOR' LOOP NEATLY
160 ON J GOTO 1000,5000,5100,2000,6000,7000
1000 PRINT CHR$(147);"UP        DOWN"
1010 PRINT"LEFT        RIGHT"
1020 PRINT"FORWARD     BACKWARD"
1030 PRINT"OPEN        CLOSE"
1040 PRINT"WRIST   -   Q"
1050 PRINT"TWIST   -   Y"
1060 PRINT"POINT   END TEACH"
1080 PRINT:PRINT NP;"POINTS"
1090 FOR I=0 TO 5:PRINT HE(I): NEXT
1100 GET A$:K=PEEK(KB):IF K=KO OR A$=""THEN 1100
1110 IF A$="E" THEN 100
1120 IF A$<>"P"THEN 1150
1130 IF NP=MP THEN GOTO 1000:REM TOO MANY POINTS
1140 NP=NP+1
1145 FOR I=0 TO 5:PT(NP,I)=HE(I):NEXT:GOTO 1000
1150 FOR I=1 TO LEN(C$)
1155 IF A$<>MID$(C$,I,1) THEN NEXT: GOTO 1000
1160 J=I: I=LEN(C$): NEXT:REM CLOSE LOOP NEATLY
1170 CH=INT((J-1)/2): DI=2*(J AND 1)-1: V=HE(CH)
1180 V=V+DI:GOSUB 9000:          REM TAKE A STEP
1190 IF PEEK(KB)=K THEN 1180:   REM KEEP STEPPING
1200 HE(CH)=V:GOTO 1000
2000 NP=0:GOTO 100
5000 IF NP=0 THEN GOTO 100:         REM NO POINTS
5010 FOR P=1 TO NP
5020 FOR I=0 TO 5: TA(I)=PT(P,I): NEXT
5030 GOSUB 8000
5040 NEXT P
5050 GOTO 100
5100 IF NP=0 THEN GOTO 100
5110 FOR P=1 TO NP:PRINT P;
5120 FOR I=0 TO 5
5130 TA(I)=PT(P,I): NEXT
5140 GOSUB 8000
5150 NEXT P
5160 GET A$:IF A$="" THEN 5110
5170 GOTO 100
6000 GOTO 100:REM PUT 'SAVE' HERE
7000 GOTO 100:REM PUT 'INPUT' HERE
8000 REM MOVE FROM HERE TO TARGET
8010 M=0: FOR CH=0 TO 5
8020 RA(CH)=ABS(TA(CH)-HE(CH))
8030 WA(CH)=SGN(TA(CH)-HE(CH))
8040 IF RA(CH)>RM THEN RM=RA(CH)
8050 NEXT: IF RM=0 THEN RETURN:      REM NO MOVE
```

```
8060 FOR CH=0 TO 5
8070 RA(CH)=RA(CH)/RM: NEXT
8100 FOR R=1 TO RM
8105 FOR CH=0 TO 5:I=RA(CH):IF I=0 THEN 8130
8110 I=I+RE(CH):IF I<1 THEN RE(CH)=I:NEXT:RETURN
8120 RE(CH)=I-1: V=HE(CH)+WA(CH): HE(CH)=V
8125 GOSUB 9000
8130 NEXT: NEXT: RETURN
9000 CO=DR(V AND 7)+CH(CH)
9010 POKE PO,CO:POKE PO,CO AND MA:POKE PO,CO
9020 RETURN
10000 DD=56579:PO=56577:KB=197:KO=64: REM CBM 64
10010 DIM DR(7): REM DRIVE VALUES WITH STROBE HI
10020 FOR I=0 TO 7: READ J: DR(I)=16*J+1: NEXT
10030 DATA 1,3,2,6,4,12,8,9: REM N-E-S-W
10040 MA=254: POKE DD,255:     REM SET TO OUTPUTS
10100 DIM CH(5),HE(5): FOR I=0 TO 5: READ J
10110 CH(I)=2*J: NEXT:          REM CHANNEL CODE
10120 DATA 1,3,5,4,2,6
10130 C$="UDLRFBOCQWTY":        REM COMMAND KEYS
10200 DIM TA(5),RA(5),WA(5),RE(5)
10210 FOR CH=0 TO 5: HE(CH)=0: RE(CH)=.5
10220 V=0: GOSUB9000: NEXT:   REM ZERO THE MOTORS
10300 NP=0:MP=20:DIM PT(5,MP)
10900 GOTO 100
```

Simple Driver and Channel Identifier

NB. Change line 10000 if using a PET.

```
10 GOTO 10000
100 INPUT"CHANNEL NUMBER,DISTANCE";CH,DI
110 FOR V=0 TO DI STEP SGN(DI)
120 GOSUB 9000:NEXT
130 GOTO 100
9000 CO=DR(V AND 7)+CH*2
9010 POKE PO,CO:POKE PO,CO AND MA:POKE PO,CO
9020 RETURN
10000 DD=56579:PO=56577:KB=197:KO=64: REM CBM 64
10010 DIM DR(7): REM DRIVE VALUES WITH STROBE HI
10020 FOR I=0 TO 7: READ J: DR(I)=16*J+1: NEXT
10030 DATA 1,3,2,6,4,12,8,9: REM N-E-S-W
10040 MA=254: POKE DD,255:     REM SET TO OUTPUTS
10900 GOTO 100
```

CHAPTER 9
Analogue Output and Position Servos

Despite the advantage of ease of interfacing, the stepper motor has no absolute position reference, and runs into more problems when speedy response is necessary. The analogue servo-motor can take its reference from a simple potentiometer, although much more sophisticated devices such as synchros and encoders can be used. The feed-back signal is subtracted from the command signal, and the difference represents the position error of the servo-motor. The servo is now driven in proportion to the error, so that it moves to reduce it. When the required position is reached, the motor ceases to require power. On the debit side, if there is a standing force load on the motor, it will have a persistent error, the value of error needed to give a drive equal to the force. To minimise this error, the servo 'loop' must be 'stiff', that is to say, a small error must give a large motor torque. This raises even more problems, since a small error can cause enough torque for the motor to pick up speed, and sprint past the target to come to rest with a larger error on the other side. Then of course it spins back again, and again... The servo has started to oscillate.

Feedback and stability

To avoid oscillation, either the stiffness must be reduced or a velocity signal must be added. A velocity term winds down the motor drive as it picks up speed, so that for any given error there is a speed at which the servo is content to freewheel. If the freewheeling speed is exceeded, the servo drive acts in reverse to slow down the motion. Thus, if the right mixture of position and velocity is made, the servo comes briskly to rest at the desired position. A velocity or 'tacho' signal can be expensive, but there are ways of obtaining a similar effect more cheaply. A certain amount of velocity feed-back comes from the motor itself, in the form of the 'back-e.m.f.' generated by the rotation of the DC motor. This limits the speed up to which the motor will run in response to a given amplifier output voltage, and adds the necessary dose of damping to a low-stiffness system. As the 'gain' of the control loop (ie the voltage out per unit of position error) is increased, so this self-damping becomes less effective. In the limit the system is 'bang-bang', driving flat out for the slightest error, and, without the addition of a velocity signal to the input of the amplifier, oscillation is inevitable.

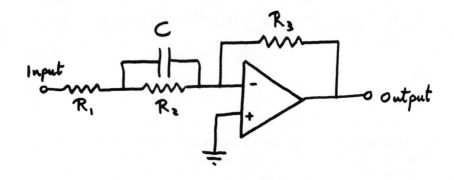

High frequency gain $= \dfrac{R_3}{R_1}$

Low frequency gain $= \dfrac{R_3}{R_1 + R_2}$

Step response (negative step input):

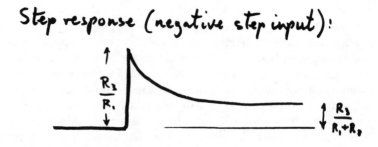

Figure 9.1   Phase advance

A high-power sophisticated servo will have a separate tacho to give a speed signal. However, this need be no more exotic than another much smaller motor connected to the main motor shaft and acting as a voltage generator. This sort of system is simple to design, and can be made extremely stiff, but for educational robots the cost of doubling-up on the motors is worth avoiding.

Phase advance

So what other possibilities are there? A substantial increase in stiffness, retaining stability, can be obtained using 'phase-advance'. The feedback resistor is split into two series resistors, and a capacitor is connected across one of these. Since the current in a capacitor is proportional to rate-of-change of voltage, the feedback signal current into a 'virtual-earth' amplifier will include a term due to rate-of-change of error position — a velocity term. Unfortunately, phase-advance also magnifies the effect of any noise on the signals, and, if the motors are driven from the same power supply that energises the position potentiometer, you will soon find oscillations aided and abetted by the power supply lines themselves.

Tacho signal from back-e.m.f.

A more cunning technique is to use the motor back-e.m.f. again, but to separate it from the voltage caused by the drive current. This involves the use of a resistance bridge and a differential amplifier — in fact, consisting of no more than three resistors to complete the bridge, and one more resistor plus a fifty pence TL081 chip. This can be very effective if enough care is taken to balance the bridge, but setting up can be fiddly.

A simple circuit

Before you leap into action and start to construct a servo system, it is only fair to warn you that the analogue output technique which follows will only work for the PET. Read to the end of the chapter, and you will find other techniques described.

For now, be content to put together a slightly less crisp servo system, which will probably do all you need. A servo module containing motor, gearbox and feedback potentiometer can be bought from a model shop for £10–£20. The motor used to try out this design was a Skyleader SRC 4BB. A drive amplifier can be made from a single-chip 759 power operational amplifier (RS number 303–258), and the power supply described in Chapter 1 can be connected to give +7V (or so) and −7V outputs. Your only problem is how to obtain an analogue drive signal from your

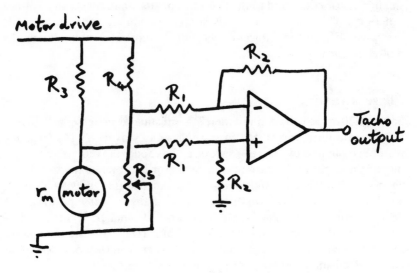

Motor drive

R_S is adjusted to make $\dfrac{R_S}{R_4} = \dfrac{r_m}{R_3}$,

where r_m is motor resistance.

R_3 is chosen to be less than r_m to avoid loss of motor power.

Output $\simeq$ $\dfrac{R_2}{R_1} \times \dfrac{R_3}{R_3 + r_m} \times$ back e.m.f.

(For $10\,\Omega$ motor, try $R_3 = 3.3\,\Omega$, $R_4 = 2.2\,k\Omega$
$R_1 = 10\,k\Omega$, $R_2 = 39\,k\Omega$, $R_5 = 1\,k\Omega$ potentiometer)

Figure 9.2   Tacho signal from motor back e.m.f.

computer. Before facing that one, get the servo loop working with a command signal taken from a second potentiometer. Connect the new potentiometer (value 1 kohm) between the 0V and +5V positions of the user port connector strip, taking the command signal from the centre wiper pin of the potentiometer. With the circuit shown below, the servo-motor should follow the command potentiometer over its full range. This circuit uses a different dodge to increase stiffness, 'integral action'. In the short term, the output voltage will be about thirty times the error voltage — this may seem ample, but note that the position error to give full drive will be 1/30 of 270 degrees, and nine degrees can be a lot of movement in a servo. The integral term means that over a second or so the drive voltage will increase to double its value, winding down any error due to a standing motor load; after ten seconds, an error of one degree can result in full torque.

Analogue output from CB2 — PET only

Now for that analogue output signal. It was mentioned in Chapter 4 that CB2 could be persuaded to act as a shift register. If control registers are set up correctly, any data byte planted in location $E84A (59466) will be repeatedly output as a serial bit pattern on CB2. Now if $E84A contains the value 1, the output pattern will be low for seven pulses and high for an eighth — the same is true for 2, 4, 8, 16, 32, 64 and 128. For value 3, the output will be low for six and high for two pulses, although 17 will give a better spacing. 73 is 01001001 in binary, and hence has three up and five down, and so on until, with 255, the output is always high. By smoothing the output with a capacitor, nine output levels can be obtained: 0, 1/8, 1/4, 3/8, 1/2, 5/8, 3/4, 7/8 and 1 times the 5V supply. Nine levels are hardly enough; how can we get better resolution? Using our old friend the binary- rate-multiplier, we can increase the resolution to 256 values.

The output is now determined by an integer, V%, say. The top three of the eight bits of V% low byte are used to select one of the eight 'patterns' 0, 1, 17, 73, 85, 182, 238 and 254 — a ninth pattern being 255. The bottom five bits of V% are masked off and added to a variable, S, say. Each time that S exceeds 31, the next higher pattern is used in the shift register, and 32 is subtracted from S. To get a smooth average, this operation must be repeated fifty or so times per second. Let's first try out the technique using a simple BASIC loop.

```
10  V% = 0: PO = 59466
20  DIM PA(8): FOR I = 0 TO 8:READ PA(I):NEXT
30  DATA 0,1,17,73,85,182,238,254,255
40  POKE 56467,16: REM MAKE CB2 A SHIFT REGISTER
50  POKE 59464,0: REM SET MAX SPEED
```

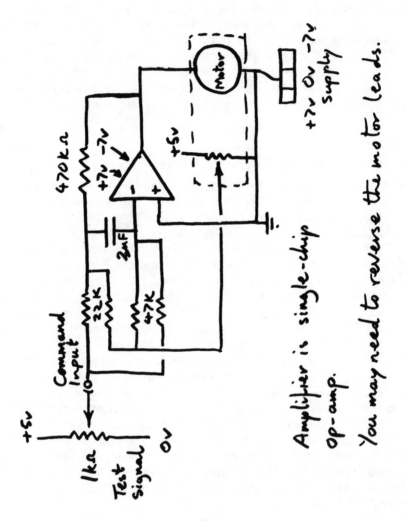

Figure 9.3   Servo amplifier

```
100  FOR V = 0 TO 255:V% = V:GOSUB 1000:NEXT
110  GOTO 100: REM OUTPUT A REPEATED RAMP

1000  A = V%/32:B = V% AND 31: REM SPLIT UP V%
1010  FOR I = 1 TO 20: REM TRY VARYING THE LOOP
1020  C = A: S = S + B
1030  IF S> 31 THEN S = S − 32:C = C + 1: REM NEXT PATTERN UP
1040  POKE PO,PA(C)
1050  NEXT I:RETURN
```

If you enter and run this program, and connect your trusty meter (6V scale) between CB2 and 0V, you should see the needle repeatedly climb slowly to 5V and then drop again. Notice that the program spends most of its time in a loop, and any time that it needs to perform a lengthy calculation, the output will take one of the nine 'steady' output values — any servo would twitch by up to thirty degrees.

Interrupt-driven output

To avoid the need for a loop in the program, the interpolation (ie averaging) can be done with an interrupt. The BASIC program goes its happy way, and, every time a certain timer gives a pulse, the computer's attention is diverted to an interrupt routine.

First this stores away the status and all necessary registers, then the processor leaps into the appropriate bit of machine code. After executing this, control must return to the tail-end of the interrupt routine which restores all the registers and status, carrying on with the original program as though nothing had happened. The interrupt routine can now look up the value of V% as saved by the BASIC program, split it and average it, outputting a new pattern every time an interrupt occurs. The output appears as if by magic, following each and every change in the value of V% with no apparent output command.

Such an interrupt occurs in the PET (and in the 64 too) sixty times per second. Its normal purpose is to scan the keyboard and update the time TI$. After saving all the registers, the machine makes an indirect jump to an address held in locations $90 and $91 (144 and 145). A new piece of software can be 'plugged in' to the interrupt routine by planting its address in these locations. At the end of the new routine, a jump must be made to the original address so that the keyboard housekeeping gets done.

Planting the new address requires rather special care: if only one of the locations has been changed when an interrupt occurs, a crash is almost inevitable. The address must therefore be planted by means of other machine-code program which temporarily inhibits interrupts. (To see the sort of thing that can happen, make sure that any useful program has been saved, then POKE 145 with any number which takes your fancy.)

The next problem is to make some space for the machine code to occupy. When the code is held as a set of data statements for planting by part of a BASIC program, the second cassette buffer will do very well (apart from being trampled by any disk commands). For something used more often, it is preferable to be able to save away both BASIC and machine-code parts in an instantly usable form. If the first line of the program is:

```
10  POKE 41,5: RUN
```

then the start of BASIC is lifted to $0501, where the main program can reside. The direct command

```
POKE 41,5: POKE 5*256,0: NEW
```

sets up the pointers to enable you to type in the rest of the program, leaving a gap of over 200 bytes in which to plant your machine code.

The first line of the upper BASIC program must be

```
10  V% = 0
```

so that V% is the first variable of the program. The machine code can now find it through the pointers at $2A, $2B (42, 43). Next comes a call SYS 1056 to the machine-code routine, which sets up the interrupt links. This swaps the address of the jump which ends the interrupt with the address already held in $90, $91. (The method works for 30xx, 40xx and 80xx series PETs). Calling the same SYS address will now swap them back to normal again.

```
 20  SYS 1056
 30  POKE 59467,16:POKE59464,0
100  INPUT "NEW VALUE (0 TO 255)"; V%
110  GOTO 100
```

Apart from the machine code, that's all there is to it! Instead of inputting values, you can output a ramp as before — no need for the POKE command though, and subroutine 1000 merely becomes a brief delay loop:

```
1000  FOR I = 1 TO 100:NEXT:RETURN
```

Otherwise, perform any sort of computation, such as:

```
100  FOR I = 1 TO 10000
110  V% = 128 + 127*SIN(I/100)
120  NEXT: GOTO 100
```

and just by storing the result in V% you will cause it to be output.

Watch out for the GOTCHA. Running the program a second time will swap the links back and kill the output routine. A third time will set them up again, and so on. Make sure that they are back to normal before editing the program, or especially before loading another program. Interrupts can be a short cut to a crash.

Something is still missing, of course: we haven't yet loaded the machine code. This is most easily done using the PET's built-in monitor. To get to the monitor, type

SYS 1024

The machine will reply with B* and two rows of characters telling you the status of the machine. Now display the memory where you want to plant the code by typing

.M 0420 0460

The screen will show a display similar to that printed below, but with different numbers to the right of the addresses. Simply drive the cursor up to each line, typing in the values shown here in place of the values on your machine, and being sure to press RETURN on each line you have altered.

```
.M  0420   0460
.:   0420   78   AD   5B   04   A6   90   85   90
.:   0428   8E   5B   04   AD   5C   04   A6   91
.:   0430   85   91   8E   5C   04   58   60   EA
.:   0438   A0   03   B1   2A   4A   4A   4A   4A
.:   0440   4A   AA   B1   2A   29   1F   18   6D
.:   0448   5D   04   C9   20   30   03   E8   29
.:   0450   1F   8D   5D   04   BD   5E   04   8D
.:   0458   4A   E8   4C   38   04   00   00   01
.:   0460   11   49   55   B6   EE   FE   FF   FF
```

When you have finished, type

.M 0420 0460

again, and check that the display is as shown here. Finally type

.X

to return to BASIC.

To save the program, machine code and all, type

POKE 41,4

This sets the start-of-BASIC pointer to normal, and you can save the program on cassette or tape in the normal way.

Whilst you can type in the machine code without needing to understand it, I am sure that you will prefer to see what it consists of. The PLANT routine starts at $0420 — hence the SYS 1056.

0420	PLANT	SEI		;INHIBIT INTERRUPTS
0421		LDA	J + 1	;SWAP JUMP POINTERS
0424		LDX	$90	
0426		STA	$90	
0428		STX	J + 1	
042B		LDA	J + 2	
042E		LDX	$91	
0430		STA	$91	
0432		STX	J + 2	
0435		CLI		;PERMIT INTERRUPTS
0436		RTS		;END OF PLANT, RETURN
0438	DTOA	LDY	#3	;READY FOR V%
043A		LDA	($2A),Y	;LOAD VALUE OF V%
043C		LSR	A	
043D		LSR	A	
043E		LSR	A	
043F		LSR	A	
0440		LSR	A	;DIVIDE BY 32
0441		TAX		;PARK IT
0442		LDA	($2A),Y	;GET V% AGAIN
0444		AND	#$1F	;MASK WITH 31
0446		CLC		;CLEAR CARRY
0447		ADC	S	;ADD S
044A		CMP	#$20	;OVER 31?
044C		BMI	NOMORE	;NO
044E		INX		;INCREASE OUTPUT ONE NOTCH
044F		AND	#$1F	;CHOP DOWN S
0451	NOMORE	STA	S	;SAVE NEW S
0454		LDA	PA,X	;LOAD X'TH PATTERN
0457		STA	$E84A	;PLANT IN SHIFT REGISTER
045A	J	JMP	DTOA	;NOTE: THIS IS NOBBLED BY ;'PLANT' TO POINT TO THE REST ;OF THE NORMAL ROUTINE.
045D	S	BYTE	0	
045E	PA	BYTE	0,1,17,73,85,182,238,254,255	

Driving radio-control servos

If you have ploughed through the preceding sections, you may feel that it is unfair to wait until now to tell you that there is another way to go about driving a servo. This method is a favourite of Alan Dibley, who uses it with devastating effect for building Micromice. What is more, 64 owners will be glad to learn that they can use it too — although they lack the machine-code monitor which makes loading the PET so easy.

For radio control, you can buy a servo-motor off-the-shelf complete with drive amplifier. There is no need to worry about feedback, stability or tacho signals. The motor comes with a three-wire connection: two are for power supply whilst the third is the command input.

Commands to the servo take the form of pulses every 20 milliseconds or so. The width of each pulse determines the commanded position, one milli-second giving full scale one way, varying to two milliseconds for full scale the other way. Once again we can use an interrupt to repeat the motor output, and we can again use the dodge of communicating with the interrupt routine by saving the value in V%. Now CB2 will give an output in the form of a train of pulses, and can be connected to the servo command input.

Radio-control servo program

To avoid the loading problem, let us revert to the technique of Chapter 5, in which the machine code is defined by data statements within the BASIC program. These will be planted in the cassette buffer by the loader and hex-to-decimal converter at 10000 onwards:

```
   10  V% = 0:GOTO 10000

10000  MC = 3*16^2 + 9*16 : I = MC : REM $0390
10010  READ A$:IF LEN(A$)< > 2 THEN 100 :REM DONE IF XXXX
10020  GOSUB 10100
10030  POKE I,A: PRINT I,A$,A
10040  I = I + 1: GOTO 10010

10100  A = ASC(A$) – 48 + 7*(A$> ":"): REM CONVERT FROM HEX
10110  B$ = MID$(A$,2)
10120  A = 16*A + ASC(B$) – 48 + 7*(B$> ":")
10130  RETURN
```

Now we can also simplify matters by using one of the user port data bits, rather than CB2. As before, there will be several differences between PET and 64, so both sets of data statements will be given:

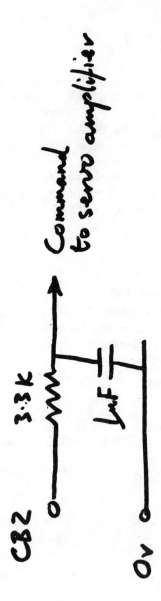

Figure 9.4 Smoothing to obtain analogue output

```
10200 DATA 78            :REM PLANT   SEI         *** PET
10210 DATA AD, CO, 03:REM             LDA J + 1
10220 DATA A6, 90        :REM         LDX $90
10230 DATA 85, 90        :REM         STA $90
10240 DATA 8E, C0, 03    :REM         STX J + 1
10250 DATA AD, C1, 03    :REM         LDA J + 2
10260 DATA A6, 91        :REM         LDX $91
10270 DATA 85, 91        :REM         STA $91
10280 DATA 8E, C1, 03    :REM         STX J + 2
10290 DATA 58            :REM         CLI
10300 DATA 60            :REM         RTS
10310 DATA A0, 03        :REM DTOA    LDY #3      V%
10320 DATA B1, 2A        :REM         LDA (VARS),Y
10330 DATA AA            :REM         TAX         – PUT IN X
10340 DATA E8            :REM         INX         – IN CASE 0
10350 DATA A9, 01        :REM         LDA #1      – SET BIT 0
10360 DATA 8D, 4F, E8    :REM         STA PORT    HIGH
10370 DATA CA            :REM LOOP    DEX         – COUNT X
10380 DATA D0, FD        :REM         BNE LOOP
10390 DATA A2, 7F        :REM         LDX #127    *VARY TO
                                                   GIVE 1 MS
10400 DATA CA            :REM LP2      DEX
10410 DATA D0, FD        :REM         BNE LP2
10420 DATA A9, 00        :REM         LDA #0
10430 DATA 8D, 4F, E8    :REM         STA PORT    – CLEAR
10440 DATA 4C, A7, 03    :REM         JMP DTOA    – GETS NOBBLED
10450 DATA XXXXX         :REM         END         BY PLANT
```

Note that the value loaded in line 10390 may need to be changed — the servo may require a different range of pulse-widths.

The CBM 64 version of the code is:

```
10200 DATA 78            :REM PLANT   SEI         *** CBM 64
10210 DATA AD, 14, 03    :REM         LDA $0314   IRQ VECTOR
10220 DATA AE, C4, 03    :REM         LDX J + 1
10230 DATA 8D, C4, 03    :REM         STA J + 1
10240 DATA 8E, 14, 03    :REM         STX $0314
10250 DATA AD, 15, 03    :REM         LDA $0315   IRQ HI
10260 DATA AE, C5, 03    :REM         LDX J + 2
10270 DATA 8D, C5, 03    :REM         STA J + 2
10280 DATA 8E, 15, 03    :REM         STX $0315
10290 DATA 58            :REM         CLI
10300 DATA 60            :REM         RTS
10310 DATA A0, 03        :REM DTOA    LDY #3      GET V%
10320 DATA B1, 2D        :REM         LDA (VARS),Y
10330 DATA AA            :REM         TAX         – PUT IN X
10340 DATA E8            :REM         INX         – IN CASE 0
10350 DATA A9, 01        :REM         LDA #1      – SET BIT 0
```

```
10360 DATA 8D, 01, DD  :REM           STA PORT    HIGH
10370 DATA CA          :REM LOOP      DEX         – COUNT X
10380 DATA D0, FD      :REM           BNE LOOP
10390 DATA A2, 7F      :REM           LDX #127    *VARY TO GIVE
                                                   1 MS
10400 DATA CA          :REM LP2       DEX
10410 DATA D0, FD      :REM           BNE LP2
10420 DATA A9, 00      :REM           LDA #0
10430 DATA 8D, 01, DD  :REM           STA PORT    – CLEAR
10440 DATA 4C, AB, 03  :REM J         JMP DTOA    – GETS NOBBLED
10450 DATA XXXXX       :REM           END         BY PLANT
```

Now we need to set bit 0 of the port as an output, and to wake up the interrupt, and we are ready to go:

```
100  POKE 59471,1  :REM IF PET
```
or
```
100  POKE 56579,1  :REM IF 64
```

```
110  SYS MC        :REM WAKE UP
```

Now the servo should respond to any value saved in variable V%, and you can use the following program as a simple test:

```
1000  FOR I = 0 TO 255
1010  V% = I: REM YES, THAT'S ALL IT TAKES
1020  FOR J = 1 TO 100 :NEXT: REM BRIEF DELAY
1030  NEXT I
1040  GOTO 1000:REM OUTPUT ANOTHER RAMP
```

When you press BREAK after running the program, the computer will carry on outputting V% on interrupt. To reset to normal, type:

SYS MC

as a direct command.

Analogue Output for PET — Basic

```
10 V%=0:PO=59466
20 DIM PA(8):FORI=0TO8:READ PA(I):NEXT
30 DATA 0,1,17,73,85,182,238,254,255
40 POKE59467,16
50 POKE59464,100
```

112

```
100 FOR V=0 TO 255:V%=V:GOSUB1000:NEXT
110 GOTO100
1000 A=V%/32:B=V%AND31
1010 FORI=1TO20
1020 C=A:S=S+B
1030 IFS>31THENS=S-32:C=C+1
1040 POKEPO,PA(C)
1050 NEXT:RETURN
```

Pulse-width Servo Output — PET

```
10 V%=0:GOTO 10000
100 POKE59471,1 :REM **** PET
110 SYS MC
120 INPUT"POSITION DEMAND (0 TO 255)";V%
130 GOTO 120
10000 MC=3*16^2+9*16:I=MC
10010 READ A$:IF LEN(A$)<>2THEN100
10020 GOSUB 10100
10030 POKE I,A:PRINT I,A$,A
10040 I=I+1:GOTO 10010
10100 A=ASC(A$)-48+7*(A$>":")
10110 B$=MID$(A$,2)
10120 A=16*A+ASC(B$)-48+7*(B$>":")
10130 RETURN
10200 DATA 78, AD,C0,03, A6,90, 85,90
10210 DATA 8E,C0,03, AD,C1,03, A6,91
10220 DATA 85,91, 8E,C1,03, 58, 60
10310 DATA A0,03, B1,2A, AA, E8, A9,01
10320 DATA 8D,4F,E8, CA, D0,FD, A2,7F
10330 DATA CA,-D0,FD, A9,00, 8D,4F,E8
10340 DATA 4C,A7,03
10350 DATA XXXX
```

Pulse-width Servo Output — CBM 64

```
10 V%=0:GOTO 10000
100 POKE56579,1 :REM **** CBM 64
110 SYS MC
120 INPUT"POSITION DEMAND (0 TO 255)";V%
130 GOTO 120
10000 MC=3*16^2+9*16:I=MC
10010 READ A$:IF LEN(A$)<>2THEN100
10020 GOSUB 10100
10030 POKE I,A:PRINT I,A$,A
10040 I=I+1:GOTO 10010
10100 A=ASC(A$)-48+7*(A$>":")
10110 B$=MID$(A$,2)
10120 A=16*A+ASC(B$)-48+7*(B$>":")
10130 RETURN
```

```
10200 DATA 78, AD,14,03, AE,C4,03
10210 DATA 8D,C4,03, 8E,14,03, AD,15,03
10220 DATA AE,C5,03, 8D,C5,03, 8E,15,03
10230 DATA 58, 60
10310 DATA A0,03, B1,2D, AA, E8, A9,01
10320 DATA 8D,01,DD, CA, D0,FD, A2,7F
10330 DATA CA, D0,FD, A9,00, 8D,01,DD
10340 DATA 4C,AB,03
10350 DATA XXXX
```

CHAPTER 10
Simple Robot Vision

Much of the work described in this chapter forms part of the research of Ali Hosseinmardi. Its publication here does not prejudice his claim to originality. The simplest vision system has been taken up by Upperdata Ltd, and is being marketed for the incredible price of £50.00. If you are industrious, you can build your own 'eye' from scratch using the information here, but you might well take the lazy way out.

Provocative instrumentation

The techniques of capturing analogue signals within the computer are many and various, and the computer is fast replacing the instrumentation recorder in process plants. Traditional instrumentation has been concerned with developing sensors which will respond to temperature, pressure, level, acidity, etc., and building them into systems which will give a steady voltage or current proportional to the quantity being measured. In many present computer applications, electrical signals which otherwise would have driven charts or pointers, are dangled under the computer's nose, to be sampled at leisure. But the computer is capable of better things than this.

The computer can perform experiments to obtain the data it needs, provoking the system to obtain a response — hence the term I have coined, provocative instrumentation. The vision system is a prime example. A sighted man will analyse the signals which happen to arrive at his eye. The blind man must tap about with his stick, building up an image of his surroundings from the responses. This image may be less detailed than the sighted one, but it is much better than no image at all.

The Cyclops vision system equips the robot with just one single focussed photocell — the blind man's stick. This one point of vision is scanned about by the robot itself, enabling a picture to be built up. The slow way is to drive the robot in a raster scan, allowing the levels of grey to be written to the display screen to build up a conventional image. More interesting is the technique of allowing the robot to follow the edge of any contrasting feature, so that the image is analysed for shape even before it is completely input. One or two cunning techniques allow features to be followed even when they are grey-on-grey, and even when the unevenness of the illumination represents more intensity variation across the field of view than the feature itself.

Making the vision system

As Mrs Beeton nearly said, 'First catch your robot.' You will need some means of deflecting the lens and photocell to scan the view. You can attach the eye to the forearm of a ready-made robot such as the Armdroid, or you can relatively easily make up a special two-channel deflection system; it could end up looking much like the home-made joystick, but with stepper motors in place of the potentiometers and the eye in place of the joystick itself. Unfortunately, the stepper motor steps are rather large, and twenty-four half-steps will cover a full ninety degrees. You will therefore need to find some way of gearing down the movement, either with conventional gears or with a simple string and pulley system. If you overdo the gearing, you can always modify the software to give you a number of steps of movement between sample points.

The eye is no more than a lens, a tube, a photocell (OP500 will do nicely) and a connecting cable. For machines similar to the PET, some extra circuitry is provided in the commercial version to give analogue-to-digital conversion (very similar to the system described in Chapter 5), but for machines with the foresight to provide a built-in converter, this is hardly necessary. In strong light, an OP500 can be connected directly to one of the paddle inputs of the 64, and for lower light levels a single transistor and two resistors can be added.

The lens should be 10 dioptres or a little stronger — ie should have a focal length of 10 cm or less. A simple plastic magnifying glass could be ideal. If you are too serious to use a Smartie tube to mount it in, a roll of paper can be held together with draughting tape. Once the photocell is fixed at the focal point and the tube is attached to the robot or scanner unit, little remains except the software. To check out your connections to the eye, and to make sure that the resistor is appropriate to the light level, you can key in and run the one-liner:

```
10  PRINT PEEK(54297):GOTO10:REM CBM 64
```

For the PET, you would have to key in line 10 and lines 10000 onwards of Chapter 5, and then test the value with:

```
100  CH% = 1:SYS MC:PRINT V%:FOR I = 1 TO 500:NEXT:
     GOTO100
```

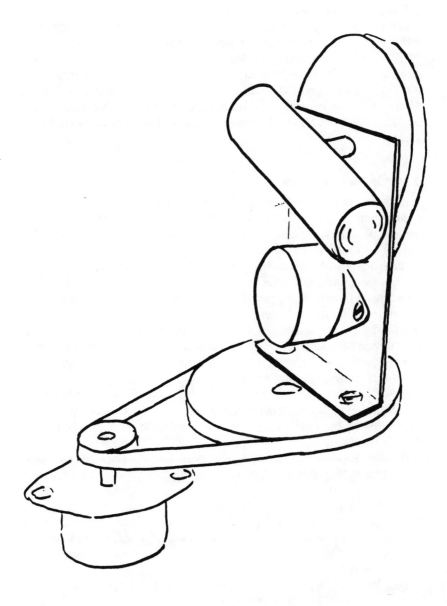

Figure 10.1   Mechanical mounting for eye

But if you use the user port for reading the photocell, where can you attach the robot? In fact robot vision is rather hard for the PET owner, not because of any problems in detecting the image but because of the lack of graphics for displaying it. If you are prepared to accept a black-and-white scene, then you can simply connect the OP500 to an LM339 comparator, set the threshold with a potentiometer, and read the light level with a single PEEK. To leave the user port clear, connect the eye to the cassette port, stealing the LM339's +5V from pin B/2. The output is now attached to pin F/6, usually used for reading the cassette switch on pin PA4 of the 6520 PIA. This port appears at address $E810, and so the input can be read by PEEK(59408) AND 16. It is this technique which is described in **Figure 10.2**.

Two vision strategies

A quick and effective way to transfer an image to the screen is by means of a raster scan. The eye is scanned to and fro, moving steadily downwards, whilst a blob, having an intensity corresponding to the photocell signal, is written on to the screen at each point. In principle, one scan line can be made from left to right, whilst the next is right to left. In practice, the backlash will cause alternate lines to be slightly shifted, and so a conventional raster is preferable. Now there are two nested FOR . . . NEXT loops to command the motor movement. The light level is read at each point, and divided by a scale factor to give a result in a range, say of 0 to 7. This number can be used to select a character from a string of increasingly dense symbols, which is then written to the corresponding screen position. Alternatively, the value can be used to select the foreground colour, and a character written which is a solid block of colour. When displayed in black and white, the right choice of colours will appear as a grey scale — you will have to use an array as a conversion table for deciding the colours, but that is easily dealt with.

An edge-following program is much more interesting. You can exploit the full resolution of the photocell, to detect edges which are light grey on dark. Let us start off by assuming that the picture is thoroughly black and white. We wish to move round the boundary of a white area, keeping black on our left and white on our right. Obviously, since we can only look at one point at a time, we will have to keep zig-zagging across the boundary to make sure that we do not lose it. A strategy that Ali and I have found effective is as follows.

Imagine that you have a compass drawn on a sheet of paper, with North and East marked. You are standing on an array of square tiles. The tile you are standing on is white, and you are facing North (as marked on the compass) along a row of more white tiles. On the paper are written the following rules:

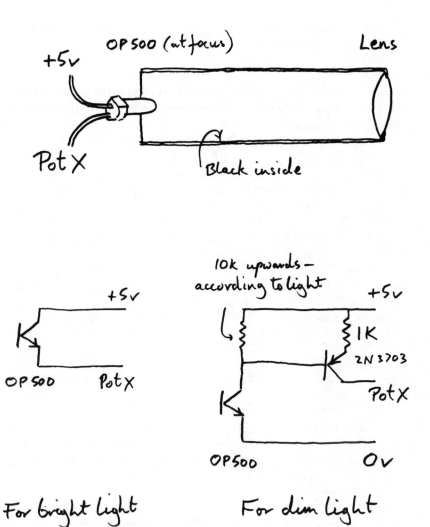

Figure 10.2   A low cost 'eye'

1.   If you are standing on a white square, rotate the compass 45 degrees to the west. Take a step 'Compass North', to the centre of the next square.

2.   If you are standing on a black square, rotate the compass 45 degrees to the east. Take a step 'Compass East', to the centre of the next square.

If you follow the rules, you will turn half left and step forwards — diagonally. If this square is black, you will turn half right (now facing your original direction) and take a sideways step to the right. You are now just one square in front of your original position. If there is a row of black squares to the left of your row of white tiles, you will zig-zag forwards advancing one square each two moves. If your white square comes to an end, you will keep on turning half right until you again reach a white square, and so you will turn any corner of the boundary. It's hard to explain, but easy to program.

Now, how can the strategy find an edge in a varying shade of grey? The technique involves setting up a local THreshold level, midway between 'Local-Black' and 'Local-White'. If the point you are now looking at is blacker than Local-Black, then Local-Black is immediately modified to that value. Similarly Local-White is immediately changed if a whiter-than-white point is found. If the level lies somewhere between these values, then both values are allowed to edge inwards slightly, by assignments such as:

LB = LB + (TH−LB)/20:REM THRESHOLD−LOCALBLACK

In this way, the black/white decision is made about a threshold which can follow the variations of illumination across the image. Ali's experimental set-up can trace the outline of a black letter, even when another sheet of paper is placed over the top of it — but the robot is apt to chase off after a crease in the paper.

By now you should have enough clues to write your own edge-following program, after you have tried out the raster program given below. It is hard to say when such a program is complete, since having captured a set of data-points representing the edge of the object, these need smoothing to remove the 'hem-stitch' pattern. They can then be processed to remove irrelevant points — straight lines can be sufficiently well represented by one point at each end, and a little cunning can reduce the outline of a K from several hundred points to fourteen. (Somehow the program seems unable to manage eleven.)

The raster program

For any strategy, it is necessary to have a procedure for moving the robot. If you are using a complete Armdroid or such, then subroutine 8000 of Chapter 8 can be called via a routine at 6000 which adds X and Y to TA(0) and TA(1), where X and Y are the displacements to be made. In addition to lines 10000 to 10900 and 8000 to 8190 of Chapter 8, you will need:

```
6000  TA(0) = TA(0) + X
6010  TA(1) = TA(1) + Y
6020  REM: THE CHANNEL NUMBERS 0,1 MAY NEED
        CHANGING – CHECK
6030  GOSUB 8000
6040  RETURN
```

If instead of a ready-built robot you are using two stepper motors of your own, you can use subroutine 8000 of Chapter 7 almost as it stands (you will need to enter lines 8000 onwards). To keep the rest of this chapter as uncomplicated as possible, add the following bodge at 6000 (you will probably later want to modify the program slightly to avoid it):

```
6000  LM = X:RM = Y:GOSUB 8000:RETURN
```

Since we might have two alternative attics to the program, let us keep the rest of the housekeeping downstairs, so that the program starts (for the 64):

```
  10  GOTO 10000
 100  LS = 16/256:EY = 54297:REM LIGHT SCALE, EYE ADDRESS
 110  CC = 55296:DIM CC(7) :REM COLOUR MAP, CODES
 120  FORI = 0 TO 7:READ CC(I):NEXT
 130  DATA 0,2,6,14,5,13,7,1
 140  REM BLACK,RED,BLUE,LT BLUE,GREEN,LT GREEN,
        YELLOW,WHITE
```

For the PET, with no shading, this is simplified to:

```
  10  GOTO 10000: REM PET ****
 100  EY = 59408:CC = 32768:REM EYE, SCREEN ADDRESS
```

continuing for both machines:

```
1000  PRINT CHR$(147);CHR$(18);:REM CLEAR SCREEN,
        REVERSE
1010  FOR RO = 0 TO 24:REM ROW
1020  FOR CM = 0 TO 39:REM COLUMN
```

```
1030 POKE     CC + 40*RO + CM,CC(PEEK(EY)*LS):REM     SET
     COLOUR *** 64 ONLY
1040 PRINT "   ";:REM PRINT (REVERSED) SPACE *** 64 ONLY
1050 X = 1:Y = 0:GOSUB 6000:REM CAN CHANGE X VALUE TO
     SCALE
1060 NEXT CM:REM NEXT COLUMN
1070 X = - 20*X:Y = - 1:GOSUB 6000:REM SCALE Y VALUE TOO
1080 NEXT RO
1090 X = 0:Y = - 25*Y:GOSUB 6000:REM MOVE BACK TO START
1100 GET A$:IF A$ = " " THEN 1100
1110 GOTO1000
```

For the PET, line 1040 is omitted and line 1030 becomes:

```
1030 POKE CC + 40*RO + CM,32 + 8*(PEEK(EY) AND 16)
```

which will plant a space if the input is low, and a reversed space if high.

This should be enough to transfer a view to the screen. If you are using a television set as a monitor, turning the colour right down will give you a grey scale. If your set is black-and-white, you win hands down!

Of course, that famous law will ensure that the photocell will start off pointing in the wrong direction. You will want to add:

```
200 PRINT CHR$(147);"SET EYE TO TOP LEFT OF SCENE"
210 INPUT "MOVE RIGHT HOW MANY? ";X
215 INPUT "MOVE DOWN HOW MANY? ";Y
220 GOSUB 6000
230 INPUT "OK?";A$:IF LEFT$(A$,1)< > "Y"THEN 200
```

As soon as you are able to capture an image in the computer, a new world opens up in which you can try edge processing, image matching, two-dimensional filtering, and a variety of advanced techniques which are the subject of current research. The computer may fall short of 'real-time' analysis by a factor of hundreds in speed, and the 'pixel' resolution may not be marvellous, but the principles of any strategy should be within the capabilities of the machine.

CHAPTER 11
Whatever Next?

It has taken a long time for the great computer manufacturers to acknow-ledge the existence of the micro. Despite their efforts at ignoring it, it wouldn't go away. At first, low-cost microcomputer systems were little more than toys, so obviously handicapped by their eight-bit inferiority that they couldn't ever threaten the mainframe — or could they? Their accessi-bility attracted many people to software writing, some of whom would cringe at the tag 'computer scientist', and, before long, a wealth of ingeni-ous packages were hitting the market, ranging from word processors to spreadsheet calculators, from payrolls to catering analysis. As the supply of peripherals moved upmarket in performance and downmarket in price, it became obvious that the micro was no mere lightweight but was set to become the cornerstone of commerce — warranting numerous commer-cials every night on television. When sixteen-bit chips appeared, the giants started to stir — although their software was mainly based on transcrip-tions of eight-bit routines and seldom gave any speed advantage. Now the mainframe manufacturers are scrambling to integrate micros into their marketing strategies.

The same story is starting to unfold for robots.

Robot evolution

Production engineers have long been familiar with numerically controlled machine tools. Controlled by barbaric means such as punched paper tape, these are identical in concept to the rest of the industrial robots. Instruc-tions programmed once are repeated to produce a stream of identical pro-ducts. Change the instructions, and the same expensive machine tool can produce a different product — the start of a flexible manufacturing system. Only when the anthropomorphic robot arm appeared did the term 'robot' gain general acceptance for this type of automation, a name carried by the IBM robot which is more like the arrangement of a milling machine than like a human arm. Industrial robots in this league carry price tags of tens of thousands of pounds, and a vision system may cost several times more.

Then the educational robots appeared on the scene. A few hundred pounds could buy a rather tinny device, admittedly resembling a toy version of the business end of a JCB. This could be connected to almost any High Street microcomputer to allow it to be programmed in a way resembling its larger cousins. Its lifting power was practically zero and its speed was not remarkable, but robots were no longer the exclusive property of major industries. Even to these humble devices, sensors could be added and linked with programs which leaned towards intelligence. As small firms (and some large institutions) experimented with the possibilities of cheap automation, it became clear that a demand was growing for the robot which could combine relative cheapness with a usable performance. Just as the microcomputer grew to fill its market place, so the microrobot is stretching its muscles to find industrial application. Unlike its upmarket forebears, the microrobot is not shackled to obsolescent and expensive minicomputers; it can exploit the latest and cheapest microcomputer technology. Complete with computer, language and performance to match today's market leaders, a new generation of robots should be priced at under £5,000.

Availability of robot-power is only half the story, as far as industrial exploitation is concerned. Few applications as yet employ a fraction of the sophistication of which the robot is capable, and a shortage of able robot programmers means that 'teach-and-repeat' programming is the most common. The end of this problem is in sight, though. With the microcomputer has grown a generation of youngsters made familiar in school and at home with micro programming. Some have made fortunes as entrepeneurs, others have found sadly that computer programming can be as lowly paid as shorthand typing. As micros throughout schools and homes become equipped with robots, so another generation will take industrial automation in its stride. Soon every back-street workshop will be able to afford a robot or two, and there will be expertise in plenty to set them up. For a while, however, experience with robots will be a highly prized commodity, and I hope that this book will give you a start.

Robot intelligence

The definition of a robot can be broadened to embrace any machine which is a 'robotnik' — a worker. It is not hard to include automatic washing machines and dish-washers, which after all measure such variables as water level and temperature and apply programmed control accordingly. Although a micromouse does little work, it is surely a robot. The mice which have struggled to the centre of the Euromicro 'Euromouse' maze have used sensors and actuators with a large amount of intelligence — if only by proxy from their designers. Industrial robots too are starting to depart from the 'do just as I tell you' image, and apply correction and adaptation to the way in which they perform their tasks, in order to achieve a more generally specified goal. (With some difficulty I suppress an urge to go into the details of the 'Craftsman Robot' project, for which my group at Portsmouth Polytechnic is receiving support from the SERC.)

The major maze-solving algorithm was established by Nick Smith, first Euromouse champion in 1980. It involved allocating numbers to the squares, starting from zero at the centre. Any square accessible from the centre was numbered one, any square accessible from a one-square was numbered two, and so on. As new walls were found, links to lower-numbered squares were broken, and the values 'floated up'. The best way to the centre was found by following the numbers downwards — until a new wall was met. A pedant would insist on calling the technique 'recursive dynamic programming'. Micromice then started to win by agility and especially reliability. David Woodfield's 'Thumper' still performs impeccably two years after its 1981 victory. Alan Dibley has introduced the concept of the 'economy mouse'. Cutting the keyboard off the cheapest available micro, he mounts it on a plywood or balsa frame controlled by commercial model aircraft servos. Using the technique mentioned in Chapter 9, he achieves a considerable measure of sucess — although not enough of late to defeat the Finnish champions. Of particular significance is the appearance in the contest of school teams, even making the pilgrimage to Madrid. Their mice may leave a lot of room for improvement, but the pattern has been set. A school team might even become the Copenhagen champions, to go on to an expenses paid trip to compete in Japan's own contest.

Robot ping-pong

Now a contest is needed which can try the mettle of the robot professionals. I have proposed a contest of robot ping-pong, and have already received several dozen serious enquiries from potential contestants (or should I say potential designers). The date for the first match was at first set for 1986, but this will now almost certainly be brought forward to 1985. The contest is not as far-fetched as it may first appear. The table is a mere half-metre wide, and is two metres long. Half-metre square frames at each end and above the net restrict the allowable movement of the ball, and reduce the area which the robots must be able to reach. The net is a quarter metre high, and this in turn makes a slam a recipe for losing a point. Simulations show that in order to make a return difficult, a robot must deliver the ball with great precision.

The serve is handled by the table itself. The ball starts at rest suspended from the centre frame above the net. When both robot vision systems have locked on to it, a nearly transparent 'fly swat' pats the ball towards the robot 'on serve'. The ball bounces once before emerging from the 'playing frame', and the robot must return it to bounce once before emerging from the opponent's playing frame. And so the game goes on.

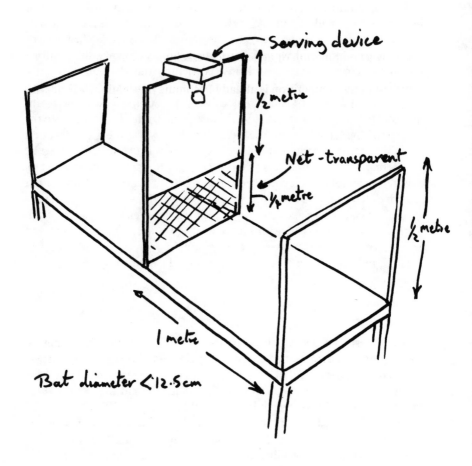

Serving device

½ metre

Net -transparent

½ metre

½ metre

1 metre

Bat diameter < 12.5 cm

Figure 11.1   Table for robot ping-pong

A few calculations show that the dexterity required is not enormous. A good X – Y plotter mounted close to the playing frame could form most of the hardware. The bat can be held by its centre by a glorified pen-lift, which is armed by a small motor and fired as the ball approaches the bat. Bat tilt can be added, or the same effect can be gained from a curved bat surface and precise placing. The number of alternative designs is at least as big as the variety of micromice, and there is no reason to suppose that entries will be confined to the 'professionals'. What is clear is that success will be earned by a combination of agile optical tracking and ingenious strategic play.

Interest in the competition is already becoming international. From the interest shown by the Japanese delegates to the Madrid Euromicro, they may be as quick to adopt robot ping-pong as they were to adopt Euromouse.

In conclusion

Microprocessors and robots may do all the pundits claim, in establishing a new industrial revolution. Manufactured goods will probably continue to slide in price, and only a nation of Luddites would continue to rely on monotonous assembly-line work as the basis of the national economy. You may be able to hasten the revolution a little; it would be hard to delay it. But whatever economic significance robots may have, they are enormous fun.

Other titles from Sunshine

SPECTRUM BOOKS

Spectrum Adventures
A guide to playing and writing adventures
Tony Bridge & Roy Carnell   £5.95
ISBN 0 946408 07 6

ZX Spectrum Astronomy
Maurice Gavin   £6.95
ISBN 0 946408 24 6

Spectrum Machine Code Applications
David Laine   £6.95
ISBN 0 946408 17 3

The Working Spectrum
David Lawrence   £5.95
ISBN 0 946408 00 9

Master your ZX Microdrive
Andrew Pennell   £6.95
ISBN 0 946408 19 X

COMMODORE 64 BOOKS

Graphic Art for the Commodore 64
Boris Allan   £5.95
ISBN 0 946408 15 7

Artificial Intelligence on the Commodore 64
Keith & Stephen Brain   £6.95
ISBN 0 946408 29 7

Commodore 64 Adventures
Mike Grace   £5.95
ISBN 0 946408 11 4

Business Applications for the Commodore 64
James Hall   £5.95
ISBN 0 946408 12 2

Mathematics on the Commodore 64
Czes Kosniowski   £5.95
ISBN 0 946408 14 9

Advanced Programming Techniques on the Commodore 64
David Lawrence   £5.95
ISBN 0 946408 23 8

The Working Commodore 64
David Lawrence   £5.95
ISBN 0 946408 02 5

Commodore 64 Machine Code Master
David Lawrence & Mark England   **£6.95**
ISBN 0 946408 05 X

ELECTRON BOOKS

Graphic Art for the Electron Computer
Boris Allan   **£5.95**
ISBN 0 946408 20 3
Programming for Education on the Electron Computer
John Scriven & Patrick Hall   **£5.95**
ISBN 0 946408 21 1

BBC COMPUTER BOOKS

Functional Forth for the BBC computer
Boris Allan   **£5.95**
ISBN 0 946408 04 1
Graphic Art for the BBC Computer
Boris Allan   **£5.95**
ISBN 0 946408 08 4
DIY Robotics and Sensors for the BBC computer
John Billingsley   **£6.95**
ISBN 0 946408 13 0
Programming for Education on the BBC computer
John Scriven & Patrick Hall   **£5.95**
ISBN 0 946408 10 6
Making Music on the BBC Computer
Ian Waugh   **£5.95**
ISBN 0 946408 26 2

DRAGON BOOKS

Advanced Sound & Graphics for the Dragon
Keith & Steven Brain   **£5.95**
ISBN 0 946408 06 8
Dragon 32 Games Master
Keith & Steven Brain   **£5.95**
ISBN 0 946408 03 3
The Working Dragon
David Lawrence   **£5.95**
ISBN 0 946408 01 7
The Dragon Trainer
A handbook for beginners
Brian Lloyd   **£5.95**
ISBN 0 946408 09 2

Sunshine also publishes

POPULAR COMPUTING WEEKLY

The first weekly magazine for home computer users. Each copy contains Top 10 charts of the best-selling software and books and up-to-the-minute details of the latest games. Other features in the magazine include regular hardware and software reviews, programming hints, computer swap, adventure corner and pages of listings for the Spectrum, Dragon, BBC, VIC 20 and 64, ZX 81 and other popular micros. Only 35p a week, a year's subscription costs £19.95 (£9.98 for six months) in the UK and £37.40 (£18.70 for six months) overseas.

DRAGON USER

The monthly magazine for all users of Dragon microcomputers. Each issue contains reviews of software and peripherals, programming advice for beginners and advanced users, program listings, a technical advisory service and all the latest news related to the Dragon. A year's subscription (12 issues) costs £10.00 in the UK and £16.00 overseas.

MICRO ADVENTURER

The monthly magazine for everyone interested in Adventure games, war gaming and simulation/role-playing games. Includes reviews of all the latest software, lists of all the software available and programming advice. A year's subscription (12 issues) costs £10 in the UK and £16 overseas.

COMMODORE HORIZONS

The monthly magazine for all users of Commodore computers. Each issue contains reviews of software and peripherals, programming advice for beginners and advanced users, program listings, a technical advisory service and all the latest news. A year's subscription costs £10 in the UK and £16 overseas.

For further information contact:
Sunshine
12–13 Little Newport Street
London WC2R 3LD
01-437 4343

Printed in England by Commercial Colour Press, London E7.